Advanced Applications of Bio-degradable Green Composites

Edited by

Amir Al-Ahmed[1] and Inamuddin[2,3,4]

[1]Centre of Research Excellence in Renewable Energy, King Fahd University of Petroleum & Minerals, Dhahran, Saudi Arabia

[2]Chemistry Department, Faculty of Science, King Abdulaziz University, Jeddah 21589, Saudi Arabia

[3]Centre of Excellence for Advanced Materials Research, King Abdulaziz University, Jeddah 21589, Saudi Arabia

[4]Department of Applied Chemistry, Faculty of Engineering and Technology, Aligarh Muslim University, Aligarh-202 002, India

Published by **Materials Research Forum LLC**
Millersville, PA 17551, USA

Published as part of the book series
Materials Research Foundations
Volume 68 (2020)
ISSN 2471-8890 (Print)
ISSN 2471-8904 (Online)

Print ISBN 978-1-64490-064-2
eBook ISBN 978-1-64490-065-9

Distributed worldwide by

Materials Research Forum LLC
105 Springdale Lane
Millersville, PA 17551
USA
https://www.mrforum.com

Manufactured in the United States of America
10 9 8 7 6 5 4 3 2 1

Table of Contents

Preface

The world community is increasingly aware and worried about non-degradable polymers used in our daily activities. Hence, bio based and biodegradable polymers and composites are of growing demand to replace petroleum based polymers and products. Every day newer materials and application concepts are publishing in many different journals. These materials are finding their applications in the medical sectors, food industry, agricultural, etc. With the progress of technologies, concepts and penetration of nanotechnology, we are expecting better performing biodegradable composites or nanocomposites for drug delivery, packaging, agricultural and many other applications. Due to the huge research activity and constant progress on biodegradable composite materials and newer application strategies, we thought to compile this book with contributions from famous researchers around the world who have agreed to share their research expertise, as well as visions for the future development of biodegradable green composites and nanocomposites.

One of the editors, Dr. Amir Al-Ahmed is thankful to the director of the Centre of Research Excellence in Renewable Energy at King Fahd University of Petroleum & Minerals, Saudi Arabia, for its continuous support. Finally, we would like to acknowledge the sincere support of Mr. Thomas Wohlbier of Materials Research Forum LLC in evolving this book into its final shape.

Amir Al-Ahmed

Centre of Research Excellence in Renewable Energy, King Fahd University of Petroleum & Minerals, Dhahran, Saudi Arabia

Inamuddin

Chemistry Department, Faculty of Science, King Abdulaziz University, Jeddah 21589, Saudi Arabia

Centre of Excellence for Advanced Materials Research, King Abdulaziz University, Jeddah 21589, Saudi Arabia

Department of Applied Chemistry, Faculty of Engineering and Technology, Aligarh Muslim University, Aligarh-202 002, India

Materials Research Forum LLC
https://doi.org/10.21741/9781644900659-1

Chapter 1

Biodegradation of Green Polymer Composites: Laboratory Procedures and Standard Test Methods

P. Rizzarelli[1*], F. Degli Innocenti[2], G. Valenti[1], M. Rapisarda[1]

[1] Istituto per i Polimeri, Compositi e Biomateriali, Consiglio Nazionale delle Ricerche, Catania, Italy

[2]Novamont S.p.A., Novara, Italy

* paola.rizzarelli@cnr.it

Abstract

Green composites have gained attention as promising alternatives to the traditional ones, mainly for their potential biodegradability. They are a combination of biodegradable polymers with natural fibres. Frequently, a compatibilization of polymeric matrix and fibre/filler for improved green composites is necessary. However, the treatments can prejudice the biodegradability of the final material. As a consequence, biodegradation tests have to been carried out. For this reason, this chapter provides an overview of both the standardization methods used to determine a degree of biodegradation in different environments and the laboratory procedures commonly adopted. Moreover, it summarizes studies in the literature, published between the beginning of 2009 and May 2019, concerning the assessment of biodegradability of green composites.

Keywords

Bioplastics, Biodegradation, Biocomposites, Standard Test Methods, Biodegradable Polymers

Contents

1.   Introduction

To create a sustainable environment and avoid the possible disposal of recalcitrant wastes, green composites have gained a lot of attention as promising alternatives to the traditional ones, particularly for their potential biodegradability [1, 2]. Typically, a green composite is a material being manufactured entirely by bio-based constituents or a mix of synthetic and bio-based constituents at a flexible percentage [3]. The first efforts towards green composites were focused on the production and characterization of systems based on recyclable polymers (e.g. polyolefins) filled with natural-organic fillers, i.e. fibres and particles obtained from plants [4]. Then, a growing worldwide ecological awareness has addressed more and more interest in composite materials from renewable, biodegradable or waste resources. Greener biocomposites from plant-derived fibre and crop-derived plastics with high bio-based content have been developed. In particular, bio-based polymers and bioplastics, as well as advanced green fibres such as lignin-based carbon fibre and nanocellulose, have been attributed great potential for sustainable composites [1, 5, 6].

Advanced Applications of Bio-degradable Green Composites   Materials Research Forum LLC
Materials Research Foundations **68** (2020) 1-44   https://doi.org/10.21741/9781644900659-1

According to the European bioplastics association, plastics that are bio-based, biodegradable, or both are known as bioplastics. The terms biodegradable, bio-based, biocompatible, compostable, etc. are currently widely used. However, these terms are still not well understood and sometimes misused. Information on this field is frequently confused, indicating that several erroneous ideas are spread both in the public opinion and in the professionals of communication. The reason of that is because the sector of biopolymers is multidisciplinary; it covers ecology, science of macromolecules, biochemistry, waste management, biodegradation testing, etc. A description based on only the viewpoint of a single discipline is not sufficient to illustrate the role of biodegradable polymer composites in the current society and in the environment.

The term "bio-based" means derived from biomass. Biomass is any material of biological origin excluding material embedded in geological formations and/or fossilized. For example, (whole or parts of) plants, trees, algae, marine organisms, microorganisms, animals, etc. [7]. Therefore, fossil oil, carbon, natural gas are not biomasses even if their origin is biological. Biomass can have undergone physical, chemical or biological treatment(s) [8]. The term bio-based is frequently used as a synonym of renewable. This will be true in most cases, but the term "renewable" material refers to a material that is composed of biomass and that can be continually replenished [9]. The biomass obtained by the exploitation of the rainforest is not "renewable" because the replenishment will be not possible in the short-medium term.

The concept of "bio-based" polymer and/or composites is strictly linked to the origin of the carbon in their backbone. The bio-based polymers derive from atmospheric carbon dioxide, which has been fixed in recent times by means of photosynthesis. On the other hand, the organic carbon of the polymers that derive from fossil resources has been trapped in soil for millions of years. As we burn fossil resources, we add to the carbon dioxide reservoir in the atmosphere. This in turn causes the greenhouse effect, i.e. an increase in the temperature of the earth's oceans and atmosphere. Polymers and materials from renewable resources are considered as beneficial for contrasting the greenhouse effect, because they recycle the atmospheric carbon [10, 11]. In biochemistry, biopolymers are polymers synthesized by living organisms. Most abundant biopolymers are polysaccharides. Cellulose and starch are polymers of glucose and are extremely abundant in the biosphere. Proteins and bacterial polyhydroxyalcanoates (PHAs) are also polymeric molecules. All these polymers are natural, i.e. they are synthesized by living organisms. Industrial exploitation is possible after extraction and purification with or without further chemical modifications. All natural unmodified bio-based polymers are biodegradable by definition. The term bio-based polymer is also applied to define polymers whose monomers derive from biomass, rather than from fossil resources. The artificial bio-based polymers can be biodegradable or not biodegradable, depending on the final chemical structure. In fact, biodegradability of any molecule is a property that depends on the chemical structure and not by the origin (fossil or biomass).

Polylactide (PLA) is a good example of bio-based polymer [12], widely studied also for blends and green composites [12-17]. Corn starch is hydrolysed to make dextrose. Dextrose is used as the fermentation feedstock and bio-converted into lactic acid. Through a special condensation process, a cyclic dimer produced from the dehydration of lactic acid (lactide) is formed. This lactide is purified through vacuum distillation and finally polymerized via ring opening polymerization. In this case, the polymer is bio-based because the original feedstock comes from the agriculture, but not natural, i.e. it is not extracted by a plant or a bacterium, but synthesized in a chemical plant by a chemist. The final polymer is biodegradable under several environmental conditions.

The cycle photosynthesis-biodegradation induces us to expect any bio-based product to be also biodegradable. This is correct as long as we consider natural materials. Things are different with artificial materials. The theorem "bio-based = biodegradable" that comes from our everyday experience does not apply here. For example, polyethylene (PE) the most common polymer in the world is made by the conventional petrochemical industry through the cracking of fossil resources. However, it can also be made starting from biomass. The bio-based PE is made starting from sugar extracted from sugar cane. Sugar is fermented into ethanol, which is converted into ethylene, which is polymerised into PE. The bio-based and the conventional PE cannot be distinguished by a chemical viewpoint. The chemical structure of PE is not attacked by microorganisms; therefore, both are not biodegradable. There is no "memory" of the "natural" origin in the bio-based PE.

Fossil-based polymers are employed in our everyday consumer goods. They are the result of the petrochemical industry and the majority is not biodegradable. However, there are fossil-based polymer that are biodegradable. Again, it is the structure and not the origin affecting the biodegradability characteristics. Polycaprolactone (PCL) is a biodegradable polyester, which can be prepared starting from crude oil to benzene, to cyclohexane, to cyclohexanone, to ε-caprolactone, and finally to the PCL. The PCL is an example of a biodegradable but not bio-based polymer [18]. Actually, it was one of the first biodegradable plastic, well studied in early 70s. PCL is known to be fast and totally biodegradable.

In development of green composites, fibre-matrix adhesion, matrix and fibre modification, and the preferred processing method are key factors in the production of high-performance materials for specific end-use applications. Frequently, a compatibilization of matrix and fibre/filler systems assembly for improved green composites is necessary and represent a key scientific challenge. In fact, biofibres are hydrophilic and hence have limited compatibility with polymer matrices, mainly hydrophobic. Several treatments (chemical, mechanical, physical, or a combination of them) have been developed to improve compatibility with the matrix. Besides the correct material selection, the final properties of composites also depend on the processing method. The design of appropriate green composites requires a clear understanding of factors influencing materials properties and performance as well as biodegradability [14].

Frequently, biodegradability and biodegradation are used wrongly as synonyms. Biodegradability refers to a property of a polymeric item, a potentiality, (i.e., the ability to be degraded by biological agents) [19]. On the other hand, biodegradation denotes a process, happening under certain conditions, in a given time, with results that can be measured. The biodegradability "status" of a polymer is deduced by studying a real biodegradation process under specific laboratory conditions and, from the test results, the conclusion that the polymer is biodegradable (i.e., it can be biodegraded) can be got.

Undoubtedly, performance as well as biodegradation of polymers, biocomposites and green composites is a topical issue [20-25]. However, the correct assessment of the biodegradability is a responsibility and represents a key scientific challenge from a combined societal, economic, environmental, and human health perspective. As a result, this chapter will deal with biodegradation of green polymer composites, reviewing standard methods and laboratory procedures. It has been divided into two parts: the first one focused on polymer biodegradation, mechanisms and test methodologies in different environments; while the second part has been dedicated to studies concerning the assessment of biodegradation of green composites, and, among these, those published between the beginning of 2009 and May 2019.

2.   Polymer biodegradation: mechanism and test methodologies in different environments

2.1   Biodegradation and biodegradability

Biodegradation is a term used in Ecology to indicate the process that brings the organic substances, produced during photosynthesis, back into inorganic substances. Biodegradation and photosynthesis function in opposite directions in the ecological cycles, most notably in the carbon biogeochemical cycle. Photosynthesis produces organic molecules starting from CO_2. Plants, algae, and all the autotrophic organisms, thanks to the sun energy, can absorb CO_2 present in the atmosphere and use it to synthesise sugars, i.e. the organic molecules at the basis of the numberless substances present in the biosphere. Through the food chain, the flow of substances and energy passes from plants (producers) to herbivores (primary consumers) and from those to carnivores (secondary consumers). On the other hand, biodegradation downgrades the organic molecules into smaller constituents and ultimately bring them back into CO_2. This final process is also called mineralization. Biodegradation is carried out by microorganisms (fungi, bacteria, protozoa), which grow on organic matter, i.e. on the refuses produced by the ecosystem. The biodegradation process is very important for the environment that must get rid of waste and residues in order to make space for new life. The ecological cycles would fast jam if it would not exist the reaction releasing CO_2 back to the atmosphere starting from the dead biomass. Therefore, the biodegradation process in the natural balance has equal dignity with the photosynthesis process of which represents the exit and in the same time the starting point. By a chemical viewpoint, biodegradation is the result of microbial respiration under aerobic conditions i.e. a chemical oxidation of organic carbon present in the substances under degradation into CO_2 and H_2O, with the concurrent releasing of energy for the cell metabolism. It is the opposite of photosynthesis, which is the reduction of CO_2 into organic carbon (i.e. carbon atom bond with other carbon atoms or hydrogen). This thermodynamically unfavourable reaction happens thanks to the solar energy.

Most polymers have a hydrophobic nature and, therefore, they are insoluble in water [26]. They are made up of macromolecules, which, because of their high molecular weight, cannot be internalized directly by microbial cells through their membrane [27, 28]. Biodegradation begins outside the microorganisms with the secretion of extracellular enzymes that affect the surface of the materials [29]. In this extracellular phase, the main reactions are hydrolysis and oxidation. Macromolecules are depolymerised into monomers and oligomers, which cross the cell membrane and join the metabolic cycle of microorganisms. Finally, the complete degradation of the metabolites involves an oxidation process and lead to production of carbon dioxide and water [30]. Biodegradation can therefore be schematically resumed in three stages (Fig.1).

The depolymerisation (Fig 1a, Stage 1) gives rise to monomers and oligomers that are incorporate by surrounding microorganisms (the first necessary condition for the biodegradation of polymers) [31]. A potential limiting factor of this phase and of the overall biodegradation process, is the solid/liquid interface available for the interaction between the microorganisms and the secreted enzymes, found in the liquid phase, and the plastic constituents in the solid one. On the other hand, the subsequent uptake of these products by microbes (Fig. 1b, Stage 2) is expected to be instantaneous [32]. Under normal conditions, Stage 3 (Fig. 1c) consists of respiration process, which leads to the mineralization of organic carbon in CO_2 and H_2O, and it is rapid at the beginning of biodegradation. Over time, microorganisms are under starvation conditions due to the lack of additional plastic to biodegrade, and products assimilated in Stage 2 (Fig. 1b) are stocked. This last phase can take a long time [33, 34].

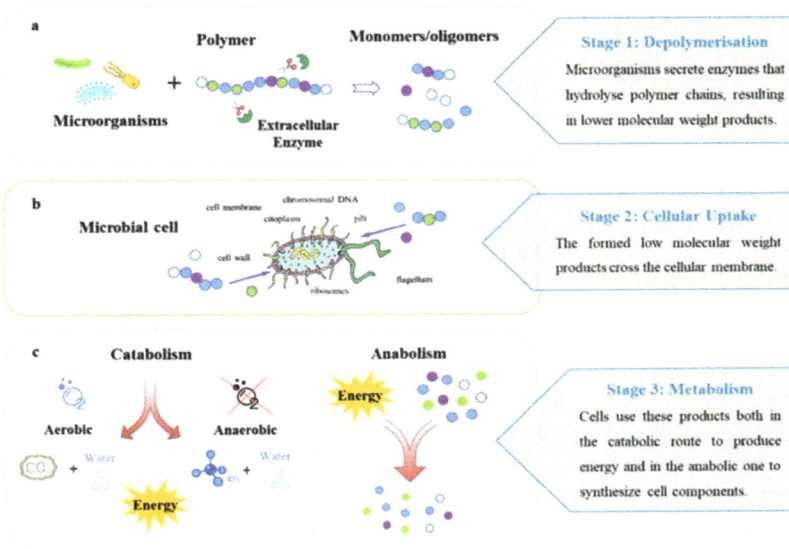

*Fig. 1Polymers biodegradation mechanism can be summarised in three stages. **(a)** In the first step, outside the cells, depolymerisation of macromolecular chains by extracellular enzymes secreted by microorganisms occurs; **(b)** then, cellular uptake of the low molecular weight degradation products takes place, followed by **(c)** their metabolisation, i.e. conversion into water, carbon dioxide, energy (catabolism) and cellular components (anabolism).*

As stated previously, biodegradability refers to a property of a polymeric item, a potentiality [19] while biodegradation refers to a process, supported by favourable conditions, within a specified time, with results that can be measured. About that it must be noted that a fully biodegradable polymer can show a very limited biodegradation if the environmental conditions are not suitable. Still, it does not lose the "status" of biodegradable. Bread roll in a freezer at $-15\ °C$ will not undergo a biodegradation process. Still the bread does not lose the status of biodegradable. It is sufficient to bring the bread back to room temperature and assure a sufficient water activity to trigger moulds and bacteria development.

Thus, first condition is the polymer must be biodegradable. Enzymes are very specific catalysts. Enzymatic reactions can only happen if the target substrate fits into the active site of the enzyme. In this way, the active moiety of the enzyme can get close enough and chemically interact with the target substrate. Synthetic molecules developed by chemists can show geometry, which do not fit with any enzyme present in the biosphere. These new molecules are therefore not biodegradable. Most traditional synthetic polymers are not biodegradable because no enzyme exists in the biosphere that recognises the new chemical "shape".

Under liquid conditions, enzyme and substrate are both in the liquid phase. Once they collide, the enzyme-substrate complex, which is the first step in the enzymatic reaction, can establish and the enzymatic reaction can happen. If the substrate is a polymeric solid material, in most cases water insoluble, biodegradation is a more complicated reaction: the enzyme is in the liquid phase while the target bond is in the solid phase. An exception to this condition is represented by the polyvinyl alcohol (PVA), a water soluble polymer. There are several factors, which can hinder the formation of the enzyme-substrate complex. Polymers are hydrophobic and this makes the access of water (and enzymes) difficult. A further factor that has been quite extensively studied is the degree of flexibility of the polymer. The polymeric chain must have enough flexibility to fits with the enzyme and allow the strict intimacy between the catalytic site and the target bond. This is not ever possible, according to the chemical structure of the polymer and the environmental temperature. Flexibility of polymers increases with the temperature. Therefore, the higher the environmental temperature the higher the polymer flexibility [35]. Additionally, the crystalline regions are less susceptible to biodegradation because more compact, more rigid and therefore more protected. Amorphous sequences are more flexible and more biodegradable [36].

This can sound trivial, but it must be mentioned. A sterile environment or with a low microbial concentration will not induce biodegradation. A proper inoculation must be assured in order to get biodegradation in laboratory testing. However, the presence of an abundant microbial population in a given environment is not a sufficient condition for the biodegradation to happen. If the population lacks those species that are endowed with the needed enzymatic activity, no biodegradation will happen. For example, no cellulose degradation will be observed in our test flask if we have inoculated a cellulase-free population. Species with the suitable genetic makeup must be present in the ecosystem (or laboratory flask) in order to get biodegradation. Most frequently, the active species are present but in minority. A strategy used by scientists working in this field is to mix inoculum withdrawn by different environments (i.e. soil, compost, activated sludge, river water, etc.) in order to increase the probability of having a very large range of microbes and enzymatic activity. Another strategy is to feed the microbial population with the material of interest before testing to allow an enrichment of the active microorganisms. The microbes that possibly can feed on the material will overgrow the others which cannot. At the moment of starting the test, the population will have a substantial fraction of microbes capable of degrading the plastic material.

Microorganisms can be present in large numbers and possess the correct genetic makeup. However, not necessarily enzymes are produced under the test conditions. Several enzymes are induced by specific metabolites. Furthermore, enzymes can be produced but be inactive, because environmental conditions are not favourable. Water, temperature, nutrients, pH, salinity etc., all these factors can affect the enzyme activity and in turn micro-organisms' growth rate. Microorganisms can be present, can have the right enzymes but biodegradation can be slow or absent, because of some environmental limiting factor.

2.2   Assessment of polymer biodegradation by standardized tests

In order to verify the biodegradability of a polymer, nowadays an experimental approach must be applied: we must follow a biodegradation process and measure what happens. The polymer biodegradation is observed under precise laboratory conditions and, from this, the conclusion that the polymer is biodegradable (specifically, it can be biodegraded) can be depicted. As described in Fig.1, three stages have been identified in biodegradation: in stage 1 the plastic (the original reactant) is depolymerised into monomers and oligomers; in stage 2 the monomers and oligomers are uptaken as biomass; in stage 3 respiration of biomass consumes O_2 and produce CO_2 and H_2O (under aerobic conditions).

The measurement of reactant consumption (i.e. the plastic material) is inconclusive, because it does not allow us to prove whether the process has actually been completed or has stopped, for example, at depolymerisation (Fig.1a, Stage 1). Measurement of biomass formation presents technical difficulties that have not been sustainably overcome to date. Therefore, all the standardized methods for determining biodegradation are based on the measurement of respiration, i.e. the conversion into CO_2 of the carbon initially present in the plastic through the use of the oxidant (O_2).

The overall reaction, which includes the three stages, can be represented as follows:

$$\text{plastic} + O_2 + \text{biomass} \rightarrow CO_2 + H_2O + \text{residual biomass.}$$

The respirometric methods differ from each other in the state (solid, liquid and sometimes biphasic), the origin of the microorganisms (the sampling environment of the microbial inoculum, for example soil, compost, etc.), the temperature and the type of measurement (disappearance of the O_2 reagent or evolution of the CO_2 product). We here consider some test methods for measuring biodegradation relevant for polymers and polymeric composites. It should be remarked that all these methods measure the degree of mineralisation, rather than the degree of biodegradation, because the production of biomass is not accounted. A reliable method for the measurement of biomass formed during a biodegradation process is still missing and thus only one product of biodegradation can be monitored. However, for reason of simplicity, this value is named degree of biodegradation.

The biodegradability of bioplastics is highly affected by several factors, among which their physical and chemical structure. On the other hand, the environment in which they are located, plays a crucial role in their biodegradation [37]. Accordingly, different test procedures in various environments, defining specific environmental conditions, have been standardised.

2.2.1 Biodegradation under controlled composting conditions

The main interest for biodegradable polymers lays on the possibility of recovery of waste by means of composting. This led to the development of a method for measuring biodegradation under specific conditions. Composting can be defined as a solid-phase, thermophilic, aerobic batch fermentation. A specific test procedure based on these conditions was early standardised in 1992 [38] and also adopted at international level as ISO 14855 [39, 40]. The polymer is mixed with mature compost and incubated at 58 °C under optimum oxygen and moisture conditions. The mature compost acts at the same time as a solid matrix and as a source of microorganisms and nutrients. The mixture is continuously aerated and the exhaust air is analysed for CO_2. The percentage of biodegradation is determined by the amount of carbon of the polymer that is converted to CO_2. Cellulose is tested in parallel as a reference material to check the activity of the inoculum. The test item is added under the form of fine powder in order to accelerate the reaction. The biodegradability test under controlled composting conditions has been standardised also by ISO in two parts [39, 40] introducing further analytical options and the possibility of using a mineral solid matrix in place of compost [41].

2.2.2 Soil biodegradation tests

About 30 % of the plastic waste produced from agriculture originates from short-life applications such as mulch films, clips, wires, nets, pheromone dispensers, geotextiles [42]. Plastic waste is frequently highly contaminated with soil or plant residues and therefore recycling can be difficult and uneconomical. Biodegradable plastics may represent a possible solution whenever retrieval from the field and recovery is difficult. Characterisation of the biodegradation behaviour in soil is important for several applications and cannot always be inferred from the biodegradation test under composting conditions. Composting is carried out at high temperature where specific microbial population grow (thermophilic) in comparison with the microbes present in soil. Furthermore, first hydrolysis for some biodegradable plastics is favoured by high temperature, which is not normally found in soil. For these reasons specific test methods based on the use of soil as the environmental matrices were developed. ISO developed in 2003 a standard test method [43] based on a respirometric approach. The polymer is introduced into soil and incubated at ambient temperature under optimum oxygen and moisture conditions. The soil acts at the same time as carrier matrix, and source of microorganisms and nutrients. Either oxygen consumption or CO_2 production is monitored.

Another relevant Standard is the ASTM D5988 [44] cited in the literature [45-47]. This test method is equivalent to ISO 17556 [43] and evaluates the degree of aerobic biodegradation of plastics exposed to soil by measuring the evolved carbon dioxide as a function of time. The rate of biodegradation is assessed in comparison with a positive reference material in an aerobic environment, thus allowing for an evaluation of the degree of biodegradability and the biodegradation time under the specific conditions. Main difference with the ISO 17556 [43] is that the ASTM standard is not based on a continuous aeration of the soil, but it is carried out within sealed vessels.

2.2.3 Marine biodegradation tests

There is a great interest towards the behaviour of biodegradable plastics when exposed to the marine environment. The reason is because the seas have been found to be heavily contaminated with non-biodegradable plastics. Biodegradable plastics are considered a possible mitigation measure for some specific applications (for example as a substitution of the plastic gears used in fish farming, quite prone to dispersion). A further series of test methods for measuring biodegradation in the marine environment has been developed more recently. Likewise, the principle is based on the exposure of the polymer to biologically active matrices withdrawn from the environments and on the measurement of respiration. The test methods ISO 19679 [48] and ISO 18830 [49] are biphasic i.e. based on a solid phase and a liquid phase. The solid phase is a sandy marine sediment laid in the bottom of a closed flask; the liquid phase is a column of seawater, poured on the sediment. The test material is preferably in the form of a film to be laid down on top of the sediment, at the interface between the solid phase and the liquid phase. This is a simulation of an object that has sunk and finally reached the sea floor. The system is contained in a closed flask. Incubation takes place preferably between 15 °C to 25 °C, but not exceeding 28 °C. In the test method ASTM D7991 [50] samples are exposed to wet marine sediments so that the microbial population is still marine but environmental conditions are different, missing here the seawater column present in the ISO standard test methods.

2.2.4 Some debated issues on the standard tests

Some issues on the standard tests have been debated. First of all, standard tests are carried out under controlled conditions in the laboratory and not in real life. In fact, the possibility to execute biodegradation test under real conditions is very difficult, almost impossible. Furthermore, by definition real conditions are not standard. Thus, the reproducibility, a pillar of the standardisation approach, which enables the possibility to verify results and claims of others (clearly a relevant point in case of certification), is lost. On the other hand, standard test methods must be validated by means of pre-normative research, i.e. there must be found evidences that the results found with the test methods are representative of what will happen on average in the real scale.

Additionally, the test material is usually tested as a powder with a high specific surface area since biodegradation only happens at the exposed surface of the polymer. Therefore, the overall biodegradation rate is very much affected by the superficial area. Testing a bulky item instead of a powder would require longer testing time, a factor that affects the steadiness of test methods. Long test methods can show drift and problems in reliability. Testing a powder is sometimes considered as a wrong way to test the polymer, because in real life bulky products undergo biodegradation. This criticism is unfounded and it is apparently based on a misunderstanding. The test of biodegradation is designed in order to measure the degree of biodegradation, relevant information needed to conclude the polymer is totally biodegradable. The focus is on understanding whether the polymer is fast and effectively biodegraded and not on the environmental fate of any specific object, whether bulky or thin.

Finally, the mineralisation is never 100 % because complete turnover of biomass into CO_2 can be very prolonged and slow especially in the last part of the biodegradation process. This happens also in natural, biodegradable polymers such as cellulose and starch [51], extensively used in green composites [52-54].

2.3   Standard specifications

Standard test methods are used to determine the degree of biodegradation or to measure other characteristics. Thus, biodegradation of any given material can range from zero to 100 %. A standard test method does not define any threshold and it is just a measurement tool. On the other hand, standard specifications specify the characteristics a material must possess in order to be designated as biodegradable. It can just be a minimum level of biodegradation or it can include other characteristics (e.g. requirements about disintegrability or about absence of ecotoxicity effects). Standard specifications also define the test methods to be used to verify the requirements. In the following paragraphs the most relevant standard specifications used in the sector of biodegradable materials are shown.

2.3.1 Organic recycling and compostability

Organic recycling is a form of recovery of biodegradable waste and bringing to compost i.e. a stabilised soil improver. Organic recycling can also include an anaerobic phase leading also to the production of biogas. The former Directive on packaging and packaging waste (94/62/EC) introduced the definition of organic recycling in 1994. Organic recycling is the aerobic (composting) or anaerobic (biomethanization) treatment by using microorganisms, under controlled conditions, of biodegradable packaging waste, which produces stabilized organic residues or methane. Landfill is not taken into account as a form of organic recycling. Packaging and plastics are in general considered as suitable to organic recycling if they comply with the relevant standard specifications. In Europe the most relevant standard is the EN 13432 [55], which is a harmonised standard being mentioned in the Official Journal of the European Communities [56]. According to the EN 13432, the characteristics of the compostable materials are:

- Biodegradability, that is the metabolic conversion of the compostable material into carbon dioxide under controlled composting conditions, according to ISO 14855 (equivalent to EN 14046 [57]). The acceptance biodegradation level is 90 % to be reached in less than 6 months.

- Disintegrability, that means fragmentation and loss of visibility in the final compost, measured in a pilot scale composting test [58]. Samples of the test material are composted with real biowaste for 3 months. The obtained compost is then screened with a 2 mm sieve. The residues of test material with dimensions > 2 mm shall be less than 10 %.

- Low levels of metals and absence of negative effects on the final compost (i.e. reduction of the agronomic value and presence of ecotoxicological effects on the plant growth). A plant growth test (modifiedOECD 208: Terrestrial Plant Test - Seedling Emergence and Seedling Growth Test) and other physical-chemical analysis are performed on compost produced starting with the test material. No difference with a control compost sample should be shown.

Each of these points is an essential requisite for the definition of compostability but it is not enough, alone. A biodegradable material is not necessarily compostable, because it must also disintegrate during the composting cycle. On the other hand, plastic items that break during composting into microscopic pieces, which are then not fully biodegradable, are also not compostable.

Other standards on organic recycling have been published, always using the same methodological approach of EN 13432 [55]. In particular, ISO 18606 [59] is the reference at International level, while ASTM D6400 [60] is the reference in North America.

2.3.2 Biodegradation in soil

A European standard on biodegradable mulch films has recently been approved [61]. The standard specifies the requirements for biodegradable films made of thermoplastic materials for use in mulching applications, which are intended to biodegrade in the ground. This standard follows the methodological approach depicted in EN 13432 [55], except that the biodegradation test is carried out on a soil sample and at a temperature suitable for activating "mesophilic" microorganisms, which are those that grow at a temperature of between 15 °C and 45 °C. In the test required by standard EN 13432 [55], the material is exposed to a sample of compost and incubated at high temperature to activate "thermophilic" microorganisms, which are those that grow at a temperature above 45 °C. The film material is considered to be biodegradable if a minimum biodegradation percentage of 90 % (absolute or relative to the reference material) is achieved in a test period not exceeding 24 months. The high biodegradation threshold is in line with the requirements of the OECD guideline and in this case is also regarded as an indicator of total biodegradation and the absence of chemical residues. On top of this another two additional stages of evaluation are needed to dissipate any doubts about the release of environmentally-hazardous molecules into the ground: a check on constituents and ecotoxicity tests. Metals subject to regulations and their maximum permissible concentrations are established on the basis of the EU criteria for the ecological quality mark for soil improvers. In addition to this, substances of very high concern are not permitted. The ecotoxicity tests are carried out in order to study possible adverse effects deriving from degradation of the mulch film in soil at the end of the specified term. After biodegradation, samples of soil incorporating high concentrations of mulch film are tested through an acute toxicity test on plants, an acute toxicity test on worms, and a test for the inhibition of nitrification by soil microorganisms.

2.4   Assessment of polymer biodegradation by laboratories procedures

Evaluation of the extent of polymer biodegradation is carried out by several methods. Different environmental conditions require diverse approaches in order to obtain valuable data on biodegradability. Eubeler et al. reviewed in 2009 validated and accepted methods based on standardized biodegradation tests, analytical assessments, enzymatic experiments or tests of physical properties to evaluate the biodegradability of synthetic polymers for different types of environments (e.g., soil, compost or aqueous media). Gravimetry, respirometry, radio-labelling, gel permeation chromatography (GPC), gas chromatography (GC) and GC mass spectrometry (GC/MS), matrix-assisted laser desorption ionization-time of flight MS (MALDI-TOF/MS) and high performance liquid chromatography MS (HPLC/MS), surface analytical methods, such as atomic force microscopy (AFM), transmission electron microscopy (TEM) and scanning electron microscopy (SEM), or nuclear magnetic resonance (NMR) and Fourier-transform infrared (FTIR) spectroscopy are all widely used methodologies [62-68]. Usually, in the same study two or more laboratory methods are applied. Same examples of more recent approaches adopted have been selected and discussed below.

Certainly, weight loss continues to be the simplest and most used method to follow polymer biodegradation. Zhuykov et al. studied the degradation of Jump to search polyhydroxybutyrate (PHB) films *in vitro* by pancreatic lipase. They based the determination of the biodegradability on the evaluation of weight loss and changes in mechanical properties [69]. Mittal et al. monitored the degradation in compost of Gum ghatti based hydrogels by weight loss at regular time interval by using:

$$\text{Weight loss (\%)} = (\text{Wi} - \text{Wr})/\text{Wi} \times 100$$

where Wi is the initial weight of the sample and Wr is the weight after the established day.

The degradation was confirmed by FTIR. A comparison of the FTIR spectra of the original and the degraded polymer showed variations in the peaks due to disintegration of the crosslinking at beginning of the degradation process. Investigations were carried out on the morphology of the polymeric samples by using SEM. Some fractures and fissures appeared at early stages of degradation, which increased in size and number with time [70].

Stable isotope analysis is a widely used technique to check the origin of organic matter in various environments. It has also been applied to detect micro traces of drugs, inflammable liquids and explosives [71]. Suzuki et al. [72] showed that carbon isotopic composition can be employed to discriminate plastic materials originated from C4 plants from those derived from petroleum. Berto et al. exploited this technique to differentiate a wide range of plastic and natural compounds. Additionally, they examined the discrepancy of $\delta^{13}C$ values in plastic items (fishing nets, passive gears, bottles caps, etc.) exposed to physical or biological phenomena occurring in the marine environment. The carbon stable isotope composition $\delta^{13}C$ of plastic samples was obtained by elemental analyser/isotope ratio MS (EA/IRMS). The isotope ratios were determined as parts per thousands (‰) using a standard reference material:

$$\delta^{13}C(‰) = [(^{13}C/^{12}C_{sample}) / (^{13}C/^{12}C_{reference}) - 1] \times 1000$$

The depletion of ^{13}C or instead an enrichment of ^{12}C observed in the "aged" plastic samples (PE and poly(ethylene terephthalate), PET, petroleum-derived polymers) was suggested to be connected to abiotic or biotic degradation in the marine environment. Similar array of degradation with a $\delta^{13}C$ increase in high density PE and biodegradable bag was detected in the experimental test. The shift of $\delta^{13}C$ was proposed to be due to physical/chemical or biological degradation. However, it was not possible to estimate the degradation rate and the process connected to the change of isotopic values [73].

Zumstein et al. issued a new approach that allowed tracking of carbon from biodegradable polymers into CO_2 and microbial biomass. It was centred on ^{13}C-labeled polymers and on isotope-specific analytical methods, involving nanoscale secondary ion MS (NanoSIMS). The Authors followed each step of biodegradation in soil of poly(butyleneadipate-co-terephthalate) (PBAT) by using different techniques. The results unequivocally demonstrated the biodegradability in soil of PBAT, a biodegradable polyester used in agriculture [74].

Reflectometric interference spectroscopy (RIfS)-based sensing system was used to evaluate the enzymatic degradation of PCL thin films. The PCL thin film degradation was simply checked by shifting the peak bottom of reflectance spectra ($\Delta\lambda$), which is proportional to the thickness of thin films. The experimental $\Delta\lambda$ values were reduced with rising the concentration of lipase from *Pseudomonas cepacia*. Attenuated total reflection FTIR spectroscopy (ATR-FTIR) imaging confirmed the method. ATR-FTIR spectra and imaging analysis on the surface of the PCL film highlighted that carbonyl groups diminished with time, as a consequence of the enzymatic hydrolysis. Moreover, disappearance of the carbonyl group was declined proportionally to the reduction in the film thickness assessed by the RIfS system [75].

3. Biodegradation of green composites

3.1 Assessment by standardized tests

The main advantage behind green composites is that they are fully biodegradable and do not have any adverse effect on the environment. As many treatments to natural materials are made, it is important to know how these modifications affect their biodegradation. Very few biodegradation studies based on the estimation of the percentage of mineralization (i.e. measuring the evolved CO_2) of the carbon content of material versus time have been performed on green composites [76-79].

In several papers, biodegradation studies are carried out according ASTM standard methods [77, 80-83]. Poly(3-hydroxybutyrate) P(3HB) and poly(3-hydroxybutyrate-co-3-hydroxyvalerate) (PHBV) are the more representative examples of PHAs, biodegradable polymers produced from a wide range of microorganisms. Wei et al. recovered and compounded low-value waste by-product from a fermentation process of potato peel waste, called the potato peel waste fermentation residue (PPW-FR) fibres, with PHB to form green composites of tuneable properties for agricultural and horticultural applications. The soil biodegradation of samples was performed according to ASTM G160 [84] and evaluated over 8 months. The surface morphology, chemistry and melting/crystallization behaviour were monitored and characterized by microscopy, FTIR spectroscopy and differential scanning calorimetry (DSC), respectively. The biocomposites showed greater biodegradation rate as compared with pure PHB, especially when the fibre content was higher than 15 %. The degradation rate increased with increasing fibre content and at 50 % of fibre content the biocomposites were completely degraded by 8 months [81]. Biocomposites of PHBV consisting of a PHBV matrix incorporating different percentage of peach palm particles (PPp), [i.e., 100/0, 90/10, 80/20 and 75/25 (% w/w) PHBV/PPp] were processed by injection moulding at 160 °C. Soil biodegradation tests were carried out under controlled temperature and moisture conditions in accordance with ASTM G160 [84]. Degradation of the biocomposites was evaluated by visual analysis, SEM and thermogravimetry analysis (TGA) up to 5 months. Soil degradation tests indicated that the biocomposites degraded faster than the neat PHBV due to the presence of holes created from the PPp introduction. Thus, degradation increased proportionally to PPp content [82]. Natural weathering and landfill burial test were carried out on PLA green composites reinforced by kenaf fibres. The rate of degradation was determined by weight loss, which was calculated once a month for six months. Natural weathering was achieved according to ASTM D1435 [85], landfill burial test was adapted from ASTM G160 [84]. The result showed that degradation was faster in a landfill burial condition being revealed a weight loss between 4 − 17 % after six months. The addition of natural fibre, whether bast or core fibre, supported the degradation of the composites increasing the weight loss rate [83].

To meet the ASTM D6400 compostability requirements [60], the constituents of plastic composites has to be tested according to ASTM D5338 [86]. The inherent biodegradation criteria of the requirements are different for diverse composition of the composites and substrates. Depending on the percentage of singular polymer or chemical modification of the natural substrates, they may need to prove conversion of 90 % of the carbon content to CO_2. In fact, crop residues, though valuable renewable resources, frequently have poor compatibility with polymers and often need modification of the biofibres to improve functional groups, thermal stability or surface area of highly reactive constituents such as cellulose, lignin and hemicelluloses. Although, ASTM D6868 [87] recognises the "materials of natural origin" as biodegradable, without further testing needs, it outlines that they must be not chemically modified and resulting from natural sources, such as wood, wood fibre, cotton fibre, starch, paper pulp or jute. However, most material will require appropriate changes to their chemical structure and composition to enable suitable adhesion. Consequently, any green composites containing chemically or biologically treated natural fibres, coproducts of agricultural material origin, need to show inherent degradability with at least 90 % of the organic carbon converted to CO_2 in 180 days at 58 °C (± 2 °C), using ASTM test method D5338 [86]. Thus, it is essential to evaluate and establish the compostability of composites containing natural substrates in order to define and design future compostable composites. As a result, a laboratory scale simulated composting facility according to ASTM D5338 was designed and employed to determine the extent of degradation of PLA and composites with modified soy straw and wheat straw as reinforcement/filler. In particular, untreated wheat and soy straw and injection moulded composites of PLA–wheat straw (70:30) and PLA–soy straw (70:30) underwent biodegradation tests. Both composite materials showed similar rates of degradation irrespective of the type of the filler used. In addition, the rate of degradation of composites was clearly higher than that of PLA (Fig. 2) [77].

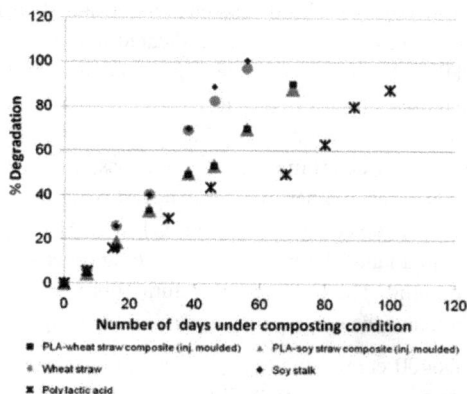

Fig. 2 Biodegradation of PLA, soy straw, wheat straw and PLA composites with wheat and soy straw under simulated aerobic compost (ASTM D 5338) [77]. (With kind permission of Elsevier)

PBAT is a synthetic polymers based on fossil resources and biodegradable. The addition of natural fillers and its effect on the final performance, including biodegradation, of the PBAT-based composites have been reported with respect to improving the properties of composites [88]. Muniyasamy et al. investigated the aerobic biodegradation of green composites from distillers dried grains with soluble (DDGS), a major co-product of the corn ethanol industry, and PBAT. Biodegradation studies were carried out according to ASTM D5338 [86]. To screen the degradation behaviour, DSC, TGA and FTIR were also employed. The experimental data showed that PBAT/treated DDGS composite was the most bio-susceptible material, being totally biodegraded. This suggested that DDGS domains were preferentially susceptible to the attack by microorganisms, and increased the percentage of biodegradation. Overall, all composites showed a degree of biodegradation similar to that of DDGS and cellulose. The Authors concluded that the incorporation of DDGS into a PBAT matrix can produce green composites with enhanced biodegradability [76].

Biodegradability evaluation of polymers and their composites in a controlled compost at 58 °C, according to ISO 14855-2 [40], was reported by Funabashi et al. PCL and PLA were employed as biodegradable polymers. Biodegradation studies were carried out in a controlled compost using a Microbial Oxidative Degradation Analyser (MODA). Noteworthy, the influence of sizes and shapes of samples, among which PLA films, cup and composites, on biodegradability was also considered. Final degrees of biodegradation and biodegradation curves for samples with various shapes showed almost the same trend [78].

The features of biodegradability, good resistance to thermal stress and worthy mechanical properties compared to common plastics have made PHBV one of the most studied biopolymers. Nevertheless, its cost and low toughness reduce its use as a substitute for traditional plastics for the production of rigid injected parts. Aiming at overcoming these limitations, novel biodegradable composites based on PHBV at various concentrations of purified α-cellulose fibres (from 3 to 45 %) were prepared by melt blending and evaluated. In a similar way, two additives with different characteristics were tested: the thermoplastic polyurethane (TPU) with the function of modifying the impact properties, and the cellulose with the function of reinforcement. Disintegrability under composting conditions was studied according to the ISO 20200 [89]. The results showed that the addition of the fibres did not affect the disintegrability of the PHBV. Therefore, the Authors established that compostable green low-cost PHBV/cellulose composites can be obtained [90, 91]. Again, Sánchez-Safont et al. suggested a formulation of a compound with a PHBV matrix to which purified α-cellulose fibres and a thermoplastic polyurethane (TPU) were added, together with three reactive agents:

hexametylenediisocianate (HMDI), a commercial multi-epoxy-functionalized styrene-co-glycidyl methacrylate oligomer (Joncryl® ADR-4368), and triglycidylisocyanurate (TGIC). This research pointed out that the reactive agents play a fundamental role in stabilizing the phases of the PHBV/TPU/cellulose compounds. Furthermore, disintegration tests in compost were carried out (ISO 20200) to highlight how the presence of reactive agents and cellulose could influence the compostability of the developed systems [89]. PHBV/30T/30C composites (30 % of TPU and 30 % of cellulose) presented different behaviour depending on the reactive agent used. The compound with TGIC and that without reactive agent were quite similar (up to 60 % weight loss); the ones with HMDI and with Joncryl® reached complete disintegration after about 70 and 50 days, respectively. This different behaviour was hypothesized to be related with a higher interaction between the cellulose and the PHBV (encouraged by the reactive agents). If the microorganisms can reach the cellulose easily the disintegration occurs first; while, if it does not happen, the TPU can be readily placed on the surface of the fibers hindering the microorganism activity [92].

Nevertheless, it has to be underlined that ISO 20200 [89] specifies a method of determining the degree of disintegration of plastic materials when exposed to a laboratory-scale composting environment. The method by itself is not applicable to the determination of the biodegradability of plastic materials under composting conditions. Further testing is required to be able to claim a material as compostable. In fact, Cinelli et al. used diverse standardized test methods to evaluate the biodegradation of composites based on PHBV and waste wood sawdust (SD) fibres, a by-product of the wood industry, produced for application in agriculture and/or plant market. The mineralization of the optimized composites was investigated under controlled composting conditions in accordance with ASTM D5338-98 (2003) [86]. The aerobic biodegradation curves recorded on the lab-scale are shown in Fig. 3.

After six months, the composite sample with the 15 wt % of SD fibres reached a mineralization percentage of about 78 %, higher than those of the control sample (filter paper) and the composite without fibres (58 %). This behaviour was explained by the presence of the fibres supporting the disintegration of the sample, increasing its susceptibility to microorganisms [79].

Additionally, disintegrability in compost was performed according to the ISO 20200 [89] and all the samples examined showed greater than 90 % disintegration. The optimized PHBV/SD composites were also used for the production of pots by injection moulding and their performance was qualitatively monitored in a plant nursery and underground for 14 months. Fig. 4 shows the results obtained with the buried pots into the soil up to the upper profile, leaving only the plants outside. Unsurprisingly, the PP pots keep on integral; those based on neat PHBV were slightly damaged on the bottom, while the pots containing the 15 wt % of SD fibres were fully fragmented and noticeably degraded, settling the results of the lab-scale degradation tests (Fig. 3) [79].

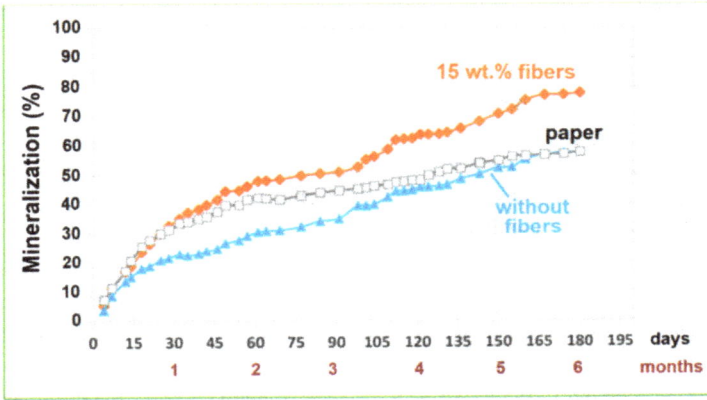

Fig. 3 Mineralization curves under simulated terrestrial environmental conditions [79]. (With kind permission of MDPI)

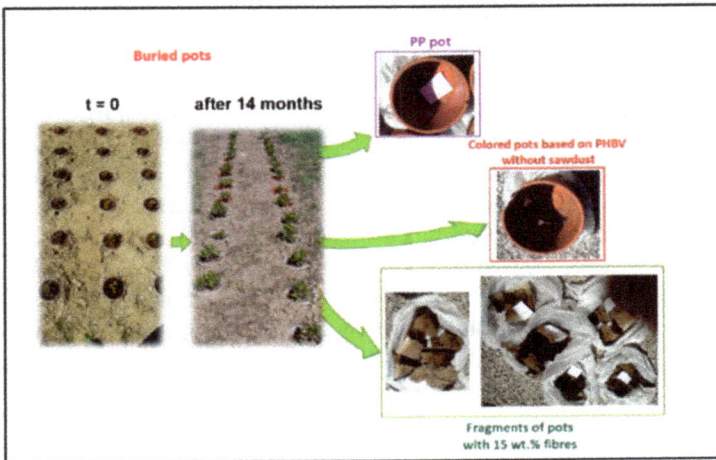

Fig. 4 Buried PP and PHB-HV based pots without fibres and with 15 wt % sawdust fibres [79]. (With kind permission of MDPI)

3.2   Assessment by laboratory procedures

In the literature, changes of a property, such as weight, molecular weight, tensile strength, tear resistance, etc. are usually measured after a certain period of incubation to monitoring biodegradation processes. Most of the papers concerning biodegradation of green composites use weight loss as a measurement of biodegradation rate [93-114]. Biodegradation studies are mainly carried out in laboratory scale and fewer in real conditions [114, 115]. It should be clear that weight loss, molecular weight changes, tensile strength decrease, etc. are only secondary results of degradation and are no proof of complete biodegradation and mineralization (Fig. 1a, Stage 1). However, they were deliberately included and discussed in this section of the chapter to give a whole overview of the studies in the literature.

Wu and co-workers carried out diverse degradation studies in soil or in compost on green composites by means of a similar approach, and monitoring weight loss, molecular weight, intrinsic viscosity and morphological changes [93-98, 116-120]. In particular, Wu investigated the structural and thermal properties of composite materials composed of PLA and green coconut fibre m(GCF). Moreover, he studied the effects of replacing pure PLA with a more compatible maleic anhydride-grafted-PLA (PLA-g-MA) in GCF-containing composites. The water absorption and weight loss of the composite films exposed to a *Burkholderiacepacia* compost were assessed in order to evaluate their water resistance and biodegradability. The bacterium completely degraded both the PLA and PLA-g-MA/GCF composite films. SEM microphotographs, taken at different incubation periods in a *B. cepacia* compost, showed severe disruption of the film structure after 9–12 days of incubation. The weight loss of PLA-g-MA/GCF (10 wt %) films increased with the GCF content. Decreases in molecular weight and intrinsic viscosity were also greater for GCF composites [93]. The biodegradable properties of composite materials made from PLA and recycled arrowroot fibre (AF), as agricultural residues, were evaluated by Wu as well. Glycidylmethacrylategrafted PLA (PLA-g-GMA) and coupling agent-treated AF (TAF) were melt blended to prepare the PLA-g-GMA/TAF composite that had enhanced properties than the PLA/AF composite. The morphological changes of its decomposition surfaces were observed by optical microscopy and by SEM. The mass losses showed that both PLA/AF and PLA-g-GMA/TAF materials underwent soil burial degradation, particularly at high levels of AF or TAF substitution (Fig. 5) [119].

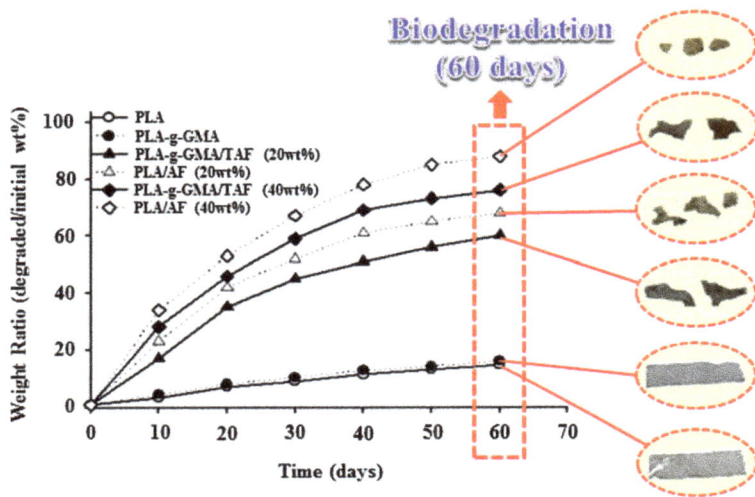

Fig. 5 Weight loss percentage and photographs of biodegraded membranes (×1 at 60 days) of PLA, PLA-g-GMA, PLA/AF, and PLA-g-GMA/TAF as a function of incubation time in soil [119]. (With kind permission of Springer Nature)

PCL is a well-known biodegradable polymer and has been proposed for use in biomedical applications and packaging. Biodegradation tests were performed on PCL and green coconut fibre (GCF) composites (PCL/GCF). Composites containing acrylic acid-grafted PCL (PCL-g-AA/GCF) were prepared as well to get better compatibility between the two components. Biodegradation tests were performed with each composite in an *Acinetobacterbaumannii* BCRC 15556 environment. Weight and mechanical properties changes were monitored to assess the progress of degradation. The mass of both composites was reduced by the GCF content within 4 weeks. The degradation rate of PCL-g-AA/GCF was only slightly lower than that of PCL/GCF [116]. In a similar way, Wu studied the biodegradability, morphology, and mechanical properties of composite materials be made of acrylic acid-grafted poly(butylene succinate adipate) (PBSA-g-AA) and agricultural residues (rice husk, RH). Composites were exposed to an *Azospirillumbrasilense* BCRC 12270 liquid culture medium that degraded both the PBSA and the PBSA-g-AA/RH films. SEM analysis pointed out erosion and cracking on the surface of the PBSA matrix and disruption of the film structure after 20-40 days of incubation. The degradation was confirmed by increasing weight loss of the PBSA matrix with incubation time (~ 16 % after 30 days). The author concluded that PBSA-g-AA/RH (20 wt %) films were not only more biodegradable than those made of PBSA but also displayed lower intrinsic viscosity and molecular weight, monitored by GPC [94].

Degradation of PHBV, modified soy protein and starch-based resins have been recently reviewed. Weight loss, mechanical property measurement, FTIR, spectroscopy, SEM, etc. have been used for monitoring the degradation behaviour [1]. The biodegradability of PHA-g-GMA/CSF composite materials made from chestnut shell fibre (CSF) and PHA, as well as CSF and glycidyl methacrylate grafted PHA (PHA-g-GMA) was evaluated. SEM, weight loss and intrinsic viscosity were used to monitoring the degradation in *Rhizobium radiobacter* compost. The Authors showed that the degradation rate of PHA-g-GMA/CSF was slower than that of PHA/CSF, but still greater than that of pure PHA. In addition, the degradation rate increased with increasing CSF content [96]. In the same way, the biodegradability, morphology, and mechanical properties of composite materials be made up of acrylic acid-grafted PHA (PHA-g-AA) and RH were investigated. Composites based on PHA-g-AA (PHA-g-AA/RH) displayed remarkably greater mechanical properties in comparison with those of PHA/RH because of greater compatibility with RH. After 60 days, the weight loss of the PHA-g-AA/RH (40 wt %) composite was greater than 90 %. PHA/RH exhibited a weight loss of ~ 4–8 wt % more than that of PHA-g-AA/RH. Both PHA/RH and PHA-g-AA/RH had higher degradation rate than pure PHA [117]. Furthermore, Wu et al. studied the biodegradability of 3D-printing filaments of composite materials made from maleic anhydride-grafted PHA (PHA-g-MA) and coupling agent-treated palm fibre (TPF). The weight loss, intrinsic viscosity and morphological changes of composite samples of PHA-g-MA and TPF (PHA-g-MA/TPF), PHA, PHA-g-MA and PHA/PF, buried in soil, were monitored over time. The extent of the morphological changes was highlighted by SEM micrographs taken after 30 and 60 days (Fig. 6). Degradation was established by the increasing weight loss of the PHA matrix as a function of incubation time, which reached about 25 % after 60 days. The Authors related the most likely reason of this weight loss to a biodegradation process. SEM micrographs showed that the PHA-g-MA/TPF (20 wt %) composites degraded faster than pure PHA.

Furthermore, the rate of weight loss of the PHA-g-MA/TPF composites was encouraged by the increase of the PF or TPF content. The degradation rate of PHA/PF was higher than that of PHA-g-MA/TPF, by about 6-9 % (Fig. 7a). It was explained with the increasing water absorption that can enlarge the surface area for microbial attack, and then leads to hydrolysis of ester groups on the PHA backbone, thus improving the rate of biodegradation. The time-course profile of the intrinsic viscosity of PHA and PHA-g-

MA/TPF incubated in soil (Fig. 7b) suggested that a higher number of fragments were present in the composites [98].

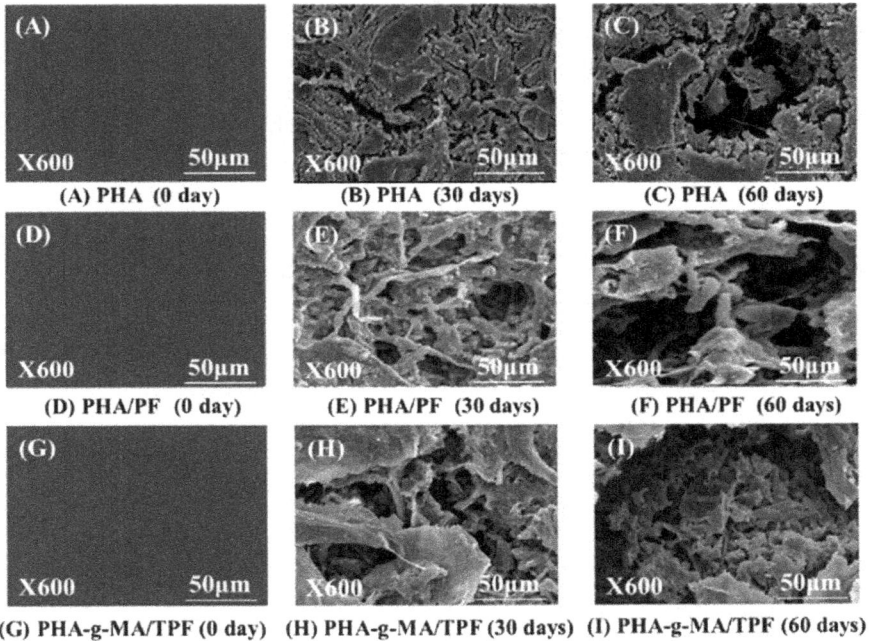

(A) PHA (0 day)   (B) PHA (30 days)   (C) PHA (60 days)

(D) PHA/PF (0 day)   (E) PHA/PF (30 days)   (F) PHA/PF (60 days)

(G) PHA-g-MA/TPF (0 day) (H) PHA-g-MA/TPF (30 days) (I) PHA-g-MA/TPF (60 days)

Fig. 6 SEM images showing the morphology of (AeC) PHA, (DeF) PHA/PF (20 wt %) and (GeI) PHA-g-MA/TPF (20 wt %) membranes as a function of incubation time in the soil [98]. (With kind permission of Elsevier)

Water absorption and soil burial tests were also performed to evaluate the durability of green composites based on bacterially synthesized poly(3-hydroxybutyrate-*co*-3-hydroxyhexanoate) [P(3HB-co-3HHx)] and short kenaf fibres (KF). Weight loss, flexural strength and molecular weight were followed during the composite degradation. The P(3HB-co-3HHx)/KF composites demonstrated higher percentage of weight loss during the soil degradation study in comparison with that of the pristine P(3HB-co-3HHx). *Burkholderia sp., Streptomyces sp., Amycolatopsis sp.* and *Streptacidiphilus sp.* were successfully isolated and characterized being involved in the degradation in vitro of the

P(3HB-co-3HHx)/KF composites during the soil degradation study [99]. Noteworthy, green composites based on PHAs and fibres of *Posidoniaoceanica* (PO) were investigated to assess their processability by extrusion, mechanical properties as well as potential biodegradability in a natural marine habitat. The degradation of composite films in a natural marine environment was evaluated in a mesocosm by weight loss measurement up to six months. More rapid degradation was recorded for films with 30 wt % PO fibres, showing that the presence of *P. oceanica* fibres assisted the disintegration of the films and, thus, speeded the degradation of the polymeric films, allowing total disintegration within six months [114].

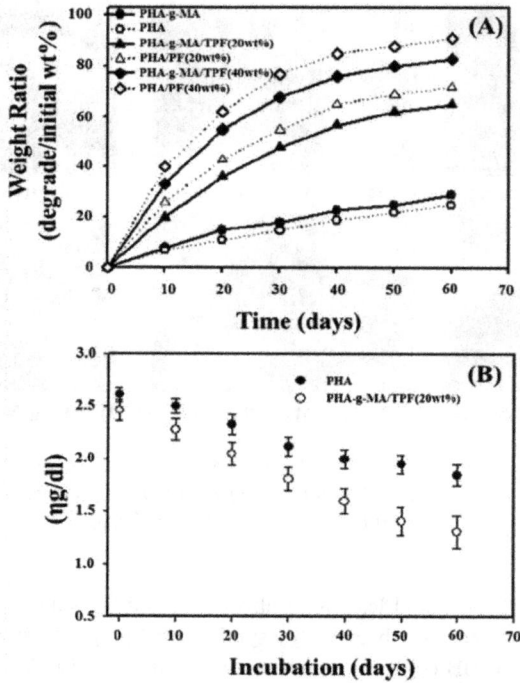

Fig. 7 (a) Weight loss percentages of PHA, PHA-g-MA, PHA/PF and PHA-g-MA/TPF as a function of incubation time in soil compost. (b) Time-course profile of the intrinsic viscosity of PHA and PHA-g-MA/TPF incubated in soil [98]. (With kind permission of Elsevier)

Advanced Applications of Bio-degradable Green Composites   Materials Research Forum LLC
Materials Research Foundations **68** (2020) 1-44   https://doi.org/10.21741/9781644900659-1

Flax fabric (FF) is one of the cheapest and strongest (high tensile and low elongation) fabrics that is made from natural flax fibres. It mainly consists of cellulose (64.1 %), hemicelluloses (16.7 %), and lignin (2 %) along with a low percentage of pectin, waxes, and inorganic materials. Compared with other natural fibres, it has low density with minimal water absorption capacity ($\sim$ 7 %). Due to its significant properties, FF-based composites have been developed and degradation in soil evaluated by weight loss percentage monitoring [121, 122]. However, a delay in soil biodegradation, measured as weight loss percentage, of biocomposites prepared by employing natural, widely available biopolymers such as thermoplastic starch (TPS), chitosan (CS) and FF has been observed [122]. On the contrary, Bayerl et al. investigated the influence of FF reinforcements on the biodegradation process of PLA in flax and PLA composites under controlled composting conditions. An experimental compost pile was set up and the specimens were exposed to the same conditions in an environment as similar as possible to a household garden compost. The progress of degradation was monitored by various quantitative and qualitative means (weight loss, morphological changes by SEM, crystallinity and Tg by DSC). Experimental data showed that fibre decomposition and hydrolysis of PLA increased the degradation rates for composites with higher short-fibre content (Fig. 8) [123].

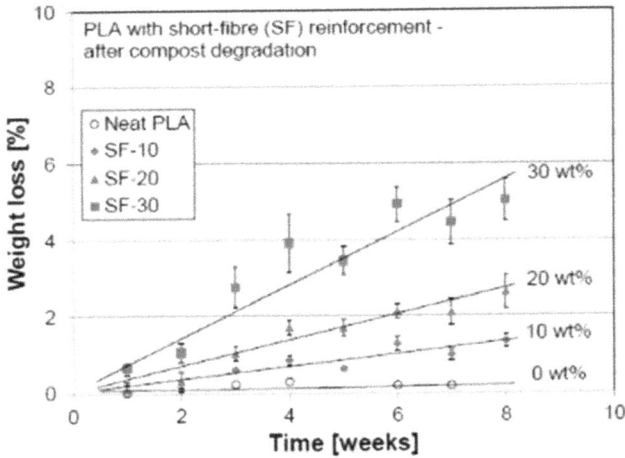

Fig. 8 Weight loss of short fibre reinforced PLA [123]. (With kind permission of Elsevier)

Thermoplastic green composites were produced by random (nonwoven mat) and aligned (unidirectional yarn) FF as reinforcements (39 % flax by volume) and PLA as matrix. Both fibre and matrix were degradable in compost soil (ready soil after composting process). Biodegradation was estimated by monitoring changes in mass, morphology (SEM analysis) and mechanical properties (flexural bending) as a function of burial time. Remarkably, the degradation property of aligned FF composites was significantly lower than that of the random ones, attributed to the less water swelling behaviour of the aligned fibre composites. After 120 days of soil burial test, the aligned flax/PLA composite displayed the decrease of 19 % mass while the random composites exhibited a loss of 27 % [106]. PLA was mixed with different amounts of wood fibres, coffee grounds, a foaming agent and fertilizer to develop a material to be used in horticulture. Biodegradation in compost was performed at 40 °C on compression moulded biocomposite samples. Weight loss was used to monitoring degradation in soil. SEM and GPC were carried out before and after biodegradation in order to evaluate changes in PLA morphology and molecular weight, respectively. Again, degradation assessment by composting tests in an aerobic environment revealed that the green composites displayed higher degradability than PLA. Biocomposites containing both wood fibres and fertilizer were the most suitable for horticulture application, combining good mechanical properties, biodegradability and fertilizer release [103]. The processability of PLA short fibre reinforced composites using three fibres (hemp, jute and lyocell) was examined. The effects of mixing zone, in the twin-screw extrusion, on the fibre lengths, mechanical properties and degradation rate of PLA natural fibre composites were evaluated. Weight loss of composite samples placed in contact with a compost medium showed that, by increasing the fibre length in PLA composites, fibre surface area within the composite was reduced and subsequently the rate of biodegradation decreased [124].

The polymers reinforced with cellulose nanofibers or with its derivatives are appealing to enhance mechanical, electrical and biodegradation properties because of the nanomeric size of the cellulose. The biodegradability, high mechanical properties, low density, and availability from renewable resources and diversity of the sources of the nanocelluloses place them as a challenging candidate for polymeric green filler [125, 126]. Fibres from the outer cell layers of the stems of various plants are called bast fibres and are mainly cellulosic in nature. Recently, Terzopoulou et al. revised the improvements on the preparation of green composites with bast fibres and aliphatic polyesters. The modifications used to increase the properties of the aliphatic polyesters in order for the final composites to achieve advanced mechanical and thermal properties as well as biodegradability were also reviewed. The authors highlighted that bastfibres enhanced the biodegradation of biocomposites, evaluated as mass loss percentage (Table 1) [102].

Table 1. Effect of bast fibres on aliphatic polyesters biodegradation [102]. (Adapted with kind permission of Elsevier)

Matrix	Fiber	Fiber content	Preparation method	Biodegradation method	Temp. (°C)	Time (days)	Mass loss (%)
PLA	Kenaf	50 vol%	Hot pressing	Composting	80	4 Weeks	38
PLA/PCL 35/15	Jute	50 wt%	Compression molding	Composting	25	15 Weeks	49
PLA	Kenaf	20 wt%	Melt mixing	Soil burial	30	3 Months	1.2
PLA	Kenaf	-	-	Enzymatic	25	6 Months	48
PBSu	Jute	10 wt%	Woven fabric	Composting	30	6 Months	62.5
PHBV	Flax	20 vol%	Compression molding	Soil burial	Environment	6 Months	20
Ecoflex1	Kenaf	10 wt%	Melt mixing	Soil burial	Environment	47 Days	2

Green composite, based on poly(butylene succinate) (PBSu) with hemp fibres and shives as fillers (from 15 to 70 wt %), were obtained by a twin screw extruder. Weight loss and crystallinity changes measured by DSC allowed to follow the degradation rate due to the action of microorganisms during soil burial test. The presence of fillers increased the rate of degradation in all the biocomposites examined, in particular in the PBSu/hemp shive composites [110]. Even the effects of crosslinking on mechanical properties and biodegradability of "protein-based" green composites were studied. A green composite was developed using wood flour and soybean protein, modified by thermal-caustic degradation and chemical crosslinking with glyoxal and polyisocyanate (PMDI). Microbial biodegradation of soybean protein-wood flour composites was investigated by SEM and weight loss. Unsurprisingly, the chemical crosslinking modification improved the mechanical properties and water resistance but decreased the biodegradability of the wood/protein composite [127]. Again, Won et al. investigated the influence of crosslinking on degradation properties in compost of kenaf/soy protein isolate- (SPI-) PVA composites. The 20 wt % of PVA and 8 wt % of glutaraldehyde (GA) produced optimum conditions for the consolidation of the composite. Experimental results confirmed that the degradation time of the composites could be controlled by the GA crosslinking agent. In particular, the degradation rate of the kenaf/SPI-PVA composite with GA was lower than that of the composite without the crosslinking agent [104].

4.   Concluding remarks

Green composites are an emerging class of biocomposites that have many potential applications to replace the traditional and synthetic one. The advantage of biodegradable green composites is the end-of-life and sustainability. However, to enhance their properties different treatments, which can prejudice the biodegradability of the final material, have been attempted. For this reason, in accordance with environmental regulations and societal concerns, biodegradation tests have to been performed.

Biodegradation can occur in different environments and to determine the biodegradation degree both standard tests and laboratory procedures have been used. The biodegradation of a polymer is observed under specific experimental conditions and, from measurable results, it can be concluded whether the polymer is biodegradable or not.

Most of the standardized methods for determining biodegradation are based on the measurement of respiration, i.e. the conversion into CO_2 of the carbon initially present in the plastic by consuming of the oxidant (O_2). The respirometric methods differ from each other in the state (solid, liquid and sometimes biphasic), the origin of microbial inoculum (for example soil, compost, etc.), the temperature and the gas monitored (disappearance of the O_2 reagent or evolution of the CO_2 produced). It should be clarified that all these methods measure the degree of mineralisation, rather than the degree of biodegradation, because the production of biomass is not accounted.

Fig. 9a shows the percentage of the papers mentioned in this chapter that adopted standardized test methods and of those concerning the analytical methods used in laboratory procedures. Remarkably, even though there are guidelines and several standardized test methods available that can be applied, in the last ten years most of the biodegradation studies on green composites have been carried out by using non standardized procedures. Fig. 9b and 9c display respectively the corresponding standard tests and analytical methods used in laboratory procedures for the biodegradation monitoring of green composites within the papers cited in the third paragraph. Overall, inspection of the literature reveals that among the standard tests the most used is ASTM D5338 for measuring biodegradation under composting conditions (Fig. 9b). In fact, one of the main advantages of biodegradable polymers is waste recovery through composting. However, in studies on the biodegradation of green composites, several works focus on laboratory soil burial ASTM G160, to test green composites for agricultural and horticultural applications.

On the contrary, the weight loss is the most widespread analytical method employed to monitoring degradation (improperly named "biodegradation"), reasonably because it is cheap, it does not require expensive equipment, and easy to perform (Fig. 9c). Usually, two or more methodologies are employed for the biodegradation tracking of green composites.

The main focus in biodegradation research has been addressed on polyesters based green composites and on soil or compost biodegradation because of the increased interest for agricultural applications or packaging materials. In spite of the great relevance, in the last ten years, biodegradation studies on green composites in aqueous media or marine environment have not been performed and published. Regardless of the approaches used, in general a higher percentage and/or rate of biodegradation or weight loss of green composites is observed, due to the presence of fibres.

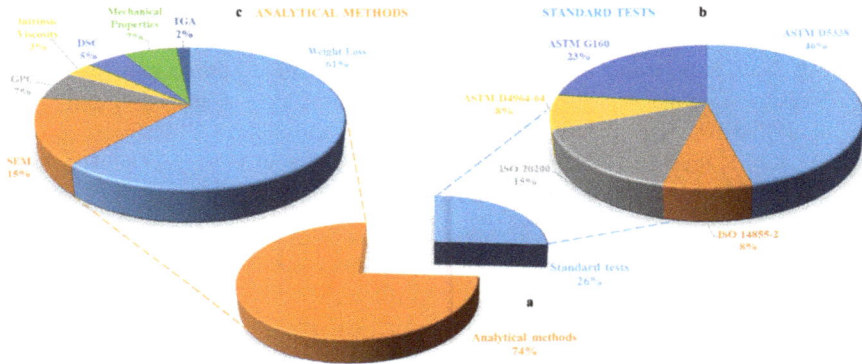

Fig. 9. (a) Percentage of the papers mentioned in this chapter that adopted standardized test methods and of those concerning the analytical ones used in laboratory procedures. Pie chart of (b) standard tests and (c) analytical methods used for the biodegradation monitoring of green composites in the papers mentioned in this chapter.

Some issues related to the use of standard methods are still subject to discussion. In fact, if on the one hand the conditions applied are not the real ones, it is nevertheless necessary to obtain reproducible results that are representative of what will happen on average in the real scale. However, research on biodegradation and conditions under which this process occurs in different environments may be relevant for the development of green composites and establish their applications with imperative impacts for society, the environment and consequently human health.

Acknowledgement

Many thanks are due to POR FSE Sicily 2020 – Project: "Polymeric systems: innovative aspects and applications in the biomedical and agri-food fields – SPIN OFF of Polymers", Call 11/2017 – "Strengthening employability in the R&D system and the emergence of research SPIN OFF in Sicily",for partial financial support.

References

[1] R. Nakamura, A.N. Netravali, Fully biodegradable "green" composites, in: T. Sabu, J. Kuruvilla, S.K. Malhotra, K. Goda, M.S. Sreekala (Eds.), Polymercomposites, Wiley-VCH Verlag GmbH & Co, 2014, pp. 431-460. https://doi.org/10.1002/9783527674220.ch12

[2] A.B. Nair, P. Sivasubramanian, P. Balakrishnan, K.A. Nair Ajith Kumar, M.S. Sreekala, Environmental effects, biodegradation, and life cycle analysis of fully biodegradable "green" composites, in: T. Sabu, J. Kuruvilla, S.K. Malhotra, K. Goda, M.S. Sreekala (Eds.), Polymercomposites, Wiley-VCH Verlag GmbH & Co, 2014, pp. 515-561. https://doi.org/10.1002/9783527674220.ch15

[3] K. Georgios, A. Silva, S Furtado, Applications of green composite materials, in: S. Kalia (Ed.), Biodegradable green composites, John Wiley & Sons, Inc., 2016, pp. 312-330. https://doi.org/10.1002/9781118911068.ch10

[4] F.P. La Mantia, M. Morreale, Green composites: A brief review, Compos. Part A Appl. Sci. Manuf. 42 (2011) 579-588. https://doi.org/10.1016/j.compositesa.2011.01.017

[5] A.K. Mohanty, S. Vivekanandhan, J.M Pin, M. Misra, Composites from renewable and sustainable resources: Challenges and innovations, Science. 362 (2018) 536-542. https://doi.org/10.1126/science.aat9072

[6] B.C. Mitra, Environment friendly composite materials: Biocomposites and green composites, Def. Sci. J. 64 (2014) 244-261. https://doi.org/10.14429/dsj.64.7323

[7] UNI EN 16575:2014, Bio-based products – Vocabulary.

[8] CEN/TR 16208:2011, Biobased products - Overview of standards.

[9] ISO 14021:2016, Environmental labels and declarations - Self-declared environmental claims (Type II environmental labelling).

[10] J. Yu, L.X.L. Chen, The greenhouse gas emissions and fossil energy requirement of bioplastics from cradle to gate of a biomass refinery, Environ. Sci. Technol. 42 (2008) 6961-6966. https://doi.org/10.1021/es7032235

[11] F.C. de Paula, C.B.C. de Paula, J. Contiero, Prospective biodegradable plastics from biomass conversion processes, in: Biofuels- State of development K. Biernat (Ed.), IntechOpen London, 2018, pp. 245-271. https://doi.org/10.5772/intechopen.75111

[12] K. M. Nampoothiri, N.R. Nair, R.P. Joh, An overview of the recent developments in polylactide (PLA) research, Bioresour. Technol. 101 (2010) 8493-8501. https://doi.org/10.1016/j.biortech.2010.05.092

[13] K. Hamada, M. Kaseemb, M. Ayyoobd, J. Joo, F. Deri, Polylactic acid blends: The future of green, light and tough, Prog. Polym. Sci. 85 (2018) 83-127. https://doi.org/10.1016/j.progpolymsci.2018.07.001

[14] T.R.K Reddy, H.J. Kim, J.W. Park, Renewable biocomposite properties and their applications, in: M. Poletto (Ed.), Composites from renewable and sustainable materials, InteTech, Croatia, 2016, pp. 177-197. https://doi.org/10.5772/65475

[15] R. Siakeng, M. Jawaid, H. Ariffin, S.M. Sapuan, M. Asim, N. Saba, Natural fiber reinforced polylactic acid composites: A Review, Polym. Compos. 40 (2019) 446-463. https://doi.org/10.1002/pc.24747

[16] P.K. Bajpai, I. Singh, J. Madaan, Development and characterization of PLA-based green composites: A review, J. Thermoplast. Compos. Mater. 27 (2014) 52-81. https://doi.org/10.1177/0892705712439571

[17] J. Sahari, S.M. Sapuan, Natural fibre reinforced biodegradable polymer composites, Rev. Adv. Mater. Sci. 30 (2011) 166-174.

[18] D.E. Perrin, J.P. English, Polycaprolactone, in: A.J. Domb, J. Kost, D.M. Wiseman (Eds.), Handbook of Biodegradable Polymers, HAP Australia, 1997, pp. 63-77.

[19] CEN/TR 15351:2006 Plastics - Guide for vocabulary in the field of degradable and biodegradable polymers and plastic items.

[20] S.K. Mary, P.K.S. Pillai, D.B. Amma, L.A. Pothen, S. Thomas, Aging and biodegradation of biocomposites, in: S. Thomas, D. Durand, C. Chassenieux, P. Jyotishkumar (Eds.), Handbook of biopolymer-based materials: from blends and composites to gels and complex network, Wiley-VCH Verlag GmbH & Co. KGaA, 2013, pp. 777-799. https://doi.org/10.1002/9783527652457.ch26

[21] D. Jayanth, P.S. Kumar, G.C. Nayak, J.S. Kumar, S.K. Pal, R. Rajasekar, A Review on biodegradable polymeric materials striving towards the attainment of green environment, J. Polym. Environ. 26 (2018) 838-865. https://doi.org/10.1007/s10924-017-0985-6

[22] S. Jayavani, H. Deka, T.O. Varghese, S.K. Nayak, Recent development and future trends in coir fiber-reinforced green polymer composites: Review and evaluation, Polym. Compos. 37 (2016) 3296-3309. https://doi.org/10.1002/pc.23529

[23] R. Siakeng, M. Jawaid, H. Ariffin, S. M. Sapuan, M. Asim, N. Saba, Natural fiber reinforced polylactic acid composites: A review, Polym. Compos. 40 (2019) 446-463. https://doi.org/10.1002/pc.24747

[24] P. Tripathi, K. Yadav, Biodegradation of natural fiber and glass fiber polymer composite-A review, Int. Res. J. Eng. Tech. 4 (2017) 1224-1228.

[25] A. Valdés, A.C. Mellinas, M. Ramos, M.C. Garrigós, Alfonso Jiménez, Natural additives and agricultural wastes in biopolymer formulations for food packaging, Front. Chem. 2 (2014) 1-10. https://doi.org/10.3389/fchem.2014.00006

[26] G.M. Bohlmann, General characteristics, processability, industrial applications and market evolution of biodegradable polymers, in: C. Bastioli (Ed.), Handbook of Biodegradable Polymers, Rapra Tech. Ltd., Shawbury, UK, 2005, pp. 183-218.

[27] R.J. Müller, Biodegradability of polymers: Regulations and methods for testing, Biopolymers Online. 10 (2005) 365-391. https://doi.org/10.1002/3527600035.bpola012

[28] F. Degli Innocenti, Biodegradable and bio-based polymers for the environment, in F. Fava, P. Canepa (Eds.), Production of fuels, specialty chemicals and biobased products from agro-industrial wastes and surplus, INCA, Venice, 2007.

[29] J. Boivin, J.W. Costerton,Biodeterioration of materials, in: H.W. Rossmoore (Ed.), Biodeterioration and biodegradation 8, Elsevier Applied Science, 1991, pp. 53-62.

[30] A.A. Shah, F. Hasan, A. Hameed, S Ahmed, Biological degradation of plastics: A comprehensive review, Biotechnol. Adv. 26 (2008) 246-265. https://doi.org/10.1016/j.biotechadv.2007.12.005

[31] D.J. Kaplan, J.M. Mayer, D. Ball, J. McMassie, A.L. Allen, P. Stenhouse, Fundamentals of biodegradable polymers, in: C. Ching, D.L. Kaplan, E.L. Thomas (Eds.), Biodegradable polymers and packaging, Technomic Pub Co, Lancaster, 1993, pp. 1-42. https://doi.org/10.1016/0306-3747(93)90255-C

[32] S. Saponaro, E. Sezenna, F. DegliInnocenti, V. Mezzanotte, L. Bonomo, A screening model for fate and transport of biodegradable polyesters in soil, J. Environ. Manage. 88 (2008) 1078-1087. https://doi.org/10.1016/j.jenvman.2007.05.010

[33] P.W. Hill, J.F. Farrar, D.L. Jones, Decoupling of microbial glucose uptake and mineralization in soil, Soil Biol. Biochem. 40 (2008) 616-624. https://doi.org/10.1016/j.soilbio.2007.09.008

[34] E. Oburger, D.L. Jones, Substrate mineralization studies in the laboratory show different microbial C partitioning dynamics than in the field, Soil Biol. Biochem. 41 (2009) 1951-1956. https://doi.org/10.1016/j.soilbio.2009.06.020

[35] R.J. Müller, E. Marten, W.D. Deckwer, Structure-biodegradability relationship of polyesters, in: E. Chiellini, H. Gil, G. Braunegg, J. Burchert, P. Gatenholm, M. van

der Zee (Eds.), Biorelated polymers - sustainable polymer science and technology, Springer - Verlag Berlin, 2001, pp. 303-311.

[36] P. Rizzarelli, C. Puglisi, G. Montaudo, Soil burial and enzymatic degradation in solution of aliphatic co-polyesters. Polym. Degrad. Stab. 85 (2004) 855-863. https://doi.org/10.1016/j.polymdegradstab.2004.03.022

[37] M. Emadian, T.T. Onay, B. Demirel, Biodegradation of bioplastics in natural environments, Waste Manag. 59 (2017) 526-536. https://doi.org/10.1016/j.wasman.2016.10.006

[38] ASTM D5338:2015, Standard test method for determining aerobic biodegradation of plastic materials under controlled composting conditions, incorporating thermophilic temperatures, ASTM International, West Conshohocken, PA, 2015

[39] ISO 14855-1:2005, Determination of the ultimate aerobic biodegradability of plastic materials under controlled composting conditions -- Method by analysis of evolved carbon dioxide- Part 1: General method.

[40] ISO 14855-2:2018, Determination of the ultimate aerobic biodegradability of plastic materials under controlled composting conditions -- Method by analysis of evolved carbon dioxide -- Part 2: Gravimetric measurement of carbon dioxide evolved in a laboratory-scale test.

[41] G. Bellia, M. Tosin, G. Floridi, F. DegliInnocenti, Activated vermiculite, a solid bed for testing biodegradability under composting conditions, Polym. Degrad. Stab. 66 (1999) 65-79. https://doi.org/10.1016/S0141-3910(99)00053-1

[42] S. Guerrini, G. Borreani, H. Vooijs, Biodegradable materials in agriculture: case histories and perspectives, in: M. Malinconico (Ed.), Soil Degradable Bioplastics for a Sustainable Modern Agriculture, Springer Book, 2017, pp. 35-65. https://doi.org/10.1007/978-3-662-54130-2_3

[43] ISO 17556, Plastics—determination of the ultimate aerobic biodegradability in soil by measuring the oxygen demand in a respirometer or the amount of carbon dioxide evolved.

[44] ASTM D5988, Standard Test Method for Determining Aerobic Biodegradation of Plastic Materials in Soil, ASTM International, West Conshohocken, PA.

[45] D. Briassoulis, A. Mistriotis, Key parameters in testing biodegradation of bio-based materials in soil, Chemosphere. 207 (2018) 18-26. https://doi.org/10.1016/j.chemosphere.2018.05.024

[46] S. Chinaglia, M. Tosin, F. Degli Innocenti, Biodegradation rate of biodegradable plastics at molecular level, Polym. Degrad. Stab. 147 (2018) 237-244. https://doi.org/10.1016/j.polymdegradstab.2017.12.011

[47] M. Tosin, A. Pischedda, F. Degli Innocenti, Biodegradation kinetics in soil of a multi-constituent biodegradable plastic, Polym. Degrad. Stab. 166 (2019) 213-218. https://doi.org/10.1016/j.polymdegradstab.2019.05.034

[48] ISO 19679, Plastics- determination of aerobic biodegradation of non-floating plastic materials in a seawater/sediment interface -- Method by analysis of evolved carbon dioxide.

[49] ISO 18830, Plastics -- Determination of aerobic biodegradation of non-floating plastic materials in a seawater/sandy sediment interface -- Method by measuring the oxygen demand in closed respirometer.

[50] ASTM D7991, Standard Test method for determining aerobic biodegradation of plastics buried in sandy marine sediment under controlled laboratory conditions, ASTM International, West Conshohocken, PA.

[51] F. DegliInnocenti, M. Tosin, C. Bastioli, Evaluation of the biodegradation of starch and cellulose under controlled composting conditions, J. Environ. Polym. Degrad. 6 (1998) 197-202. https://doi.org/10.1023/A:1021825715232

[52] V.K. Thakur, M.K. Thakur, Processing and characterization of natural cellulose fibers/thermoset polymer composites, Carbohydr. Polym. 109 (2014) 102-117. https://doi.org/10.1016/j.carbpol.2014.03.039

[53] S. Wang, A. Lu, L. Zhang, Recent advances in regenerated cellulose materials, Prog. Polym. Sci. 53 (2016) 169-206. https://doi.org/10.1016/j.progpolymsci.2015.07.003

[54] K.W. Tan, S.K. Heo, M.L. Foo, I.M.L Chew, C.K. Yoo, An insight into nanocellulose as soft condensed matter: Challenge and future prospective toward environmental sustainability, Sci. Total Environ. 650 (2019) 1309-1326. https://doi.org/10.1016/j.scitotenv.2018.08.402

[55] EN 13432:2000, Packaging - Requirements for packaging recoverable through composting and biodegradation - Test scheme and evaluation criteria for the final acceptance of packaging.

[56] Legislation 190, Official Journal of the European Communities. 44 (2001) 21-23.

[57] EN 14046, Packaging - Evaluation of the ultimate aerobic biodegradability of packaging materials under controlled composting conditions - Method by analysis of released carbon dioxide.

[58] EN 14045:2003, Packaging. Evaluation of the disintegration of packaging materials in practical oriented tests under defined composting conditions.

[59] ISO 18606:2013, Packaging and the environment - Organic recycling.

[60] ASTM D6400 Standard specification for labeling of plastics designed to be aerobically composted in municipal or industrial facilities, ASTM International, West Conshohocken, PA.

[61] EN 17033:2018 Plastics - Biodegradable mulch films for use in agriculture and horticulture - Requirements and test methods.

[62] J.P. Eubeler, M. Bernhard, S. Zok, T.P. Knepper, Environmental biodegradation of synthetic polymers I. Test methodologies and procedures, Trends Anal. Chem. 28 (2009) 1057-1072. https://doi.org/10.1016/j.trac.2009.06.007

[63] P. Rizzarelli, S. Carroccio, Modern mass spectrometry in the characterization and degradation of biodegradable polymers, Anal. Chim. Acta. 808 (2014) 18-43. https://doi.org/10.1016/j.aca.2013.11.001

[64] P. Rizzarelli, S. Carroccio, Recent trends in the structural characterization and degradation of biodegradable polymers by modern mass spectrometry, in: C.C. Chu (Ed.), Biodegradable polymers volume 1: Advancement in biodegradation study and applications, NOVA Science Publisher Inc., 2015.

[65] K. Fukushima, D. Tabuani, C. Abbate, M. Arena, P. Rizzarelli, Preparation, characterization and biodegradation of biopolymer nanocomposites based on fumed silica, Eur. Polym. J. 47 (2011) 139-152. https://doi.org/10.1016/j.eurpolymj.2010.10.027

[66] P. Rizzarelli, M. Cirica, G. Pastorelli, C. Puglisi, G. Valenti, Aliphatic poly(ester amide)s from sebacic acid and aminoalcohols of different chain length: Synthesis, characterization and soil burial degradation, Polym. Degrad. Stab. 121 (2015) 90-99. https://doi.org/10.1016/j.polymdegradstab.2015.08.010

[67] E. Castro-Aguirre, R. Auras, S. Selke, M. Rubino, T. Marsh, Insights on the aerobic biodegradation of polymers by analysis of evolved carbon dioxide in simulated composting conditions, Polym. Degrad. Stab. 137 (2017) 251-271. https://doi.org/10.1016/j.polymdegradstab.2017.01.017

[68] M. Siotto, L. Zoia, M. Tosin, F. DegliInnocenti, M. Orlandi, V. Mezzanotte, Monitoring biodegradation of poly(butylene sebacate) by gel permeation chromatography, ^{1}H-NMR and ^{31}P-NMR techniques, J. Environ. Manage. 116 (2013) 27-35. https://doi.org/10.1016/j.jenvman.2012.11.043

[69] V. Zhuikov, A. Bonartsev, D. Bagrov, A. Rusakov, A. Useinov, V. Myshkina, T. Mahina, K. Shaitan, G. Bonartseva, The changes in surface morphology and mechanical properties of poly(3-Hydroxybutyrate) and its copolymer films during in vitro degradation, Solid State Phenom. 258 (2017) 354-357. https://doi.org/10.4028/www.scientific.net/SSP.258.354

[70] H. Mittal, SB. Mishra, AK. Mishra, BS. Kaith, R. Jindal, Flocculation characteristics and biodegradation studies of gum ghatti based hydrogels, Int. J. Biol. Macromol. 58 (2013) 37-46. https://doi.org/10.1016/j.ijbiomac.2013.03.045

[71] W. Meier-Augenstein, Forensic isotope analysis, in: McGraw-Hill (Ed.), Yearbook of science & technology, McGraw-Hill, 2014, pp. 120

[72] Y. Suzuki, F. Akamatsu, R. Nakashita, T. Korenaga, A novel method to discriminate between pant- and petroleum-derived plastics by stable isotope analysis, Chem. Lett. 39 (2010) 998-999. https://doi.org/10.1246/cl.2010.998

[73] D. Berto, F. Rampazzo, C. Gion, S. Noventa, F. Ronchi, U. Traldi, G. Giorgi, A.M. Cicero, O. Giovanardi, Preliminary study to characterize plastic polymers using elemental analyser/isotope ratio mass spectrometry (EA/IRMS), Chemosphere. 176 (2017) 47-56. https://doi.org/10.1016/j.chemosphere.2017.02.090

[74] M.T. Zumstein, A. Schintlmeister, T.F. Nelson, R. Baumgartner, D. Woebken, M. Wagner, H.E. Kohler, K. McNeill, M. Sander, Biodegradation of synthetic polymers in soils: Tracking carbon into CO_2 and microbial biomass, Science Advances. 4 (2018) 9024. https://doi.org/10.1126/sciadv.aas9024

[75] T. Ooya, Y. Sakata, H.W. Choi, T. Takeuchi, Reflectometric interference spectroscopy-based sensing for evaluating biodegradability of polymeric thin films, Acta Biomater. 38 (2016) 163-167. https://doi.org/10.1016/j.actbio.2016.04.022

[76] S. Muniyasamy, M.M. Reddy, M. Misra, A. Mohanty, Biodegradable green composites from bioethanol co-product and poly(butylene adipate-co-terephthalate), Ind. Crops Prod. 43 (2013) 812-819. https://doi.org/10.1016/j.indcrop.2012.08.031

[77] R. Pradhan, M. Misra, L. Erickson, A. Mohanty, Compostability and biodegradation study of PLA–wheat straw and PLA–soy straw based green composites in simulated composting bioreactor, Bioresour. Technol. 101 (2010) 8489-8491. https://doi.org/10.1016/j.biortech.2010.06.053

[78] M. Funabashi, F. Ninomiya, M. Kunioka, Biodegradability evaluation of polymers by ISO 14855-2, Int. J. Mol. Sci. 10 (2009) 3635-3654. https://doi.org/10.3390/ijms10083635

[79] P. Cinelli, M. Seggiani, N. Mallegni, V. Gigante, A. Lazzeri, Processability and degradability of PHA-Based composites in terrestrial environments, Int. J. Mol. Sci. 20 (2019) 284-297. https://doi.org/10.3390/ijms20020284

[80] E. Sun, G. Liao, Q. Zhang, P. Qu, G. Wu, Y. Xu, Cheng Yong, H. Huang, Green preparation of straw fiber reinforced hydrolyzed soy protein isolate/urea/formaldehyde composites for biocomposite flower pots application, Materials. 11 (2018)1695-1708. https://doi.org/10.3390/ma11091695

[81] L. Wei, S. Liang, A.G. McDonald, Thermophysical properties and biodegradation behavior of green composites made from polyhydroxybutyrate and potato peel waste fermentation residue, Ind. Crops Prod. 69 (2015) 91-103. https://doi.org/10.1016/j.indcrop.2015.02.011

[82] K.C. Batista, D.A.K. Silva, L.A.F. Coelho, S.H. Pezzin, A.P.T. Pezzin, Soil biodegradation of PHBV/peach palm particles biocomposites, J. Polym. Environ. 18 (2010) 346-354. https://doi.org/10.1007/s10924-010-0238-4

[83] S.N. Surip, W.N. Raihan, W. Jaafar, Comparison study of the bio-degradation property of polylactic acid (PLA) green composites reinforced by kenaffibers, Int. J. Tech. 6 (2018) 1205-1215. https://doi.org/10.14716/ijtech.v9i6.2357

[84] ASTM G160, Standard practice for evaluating microbial susceptibility of non metallic materials by laboratory soil burial, ASTM International, West Conshohocken, PA.

[85] ASTM D1435, Standard practice for outdoor weathering of plastics, ASTM International, West Conshohocken, PA.

[86] ASTM D5338-98, Standard test method for determining aerobic biodegradation of plastic materials under controlled composting conditions, ASTM International, West Conshohocken, PA, 2003.

[87] ASTM D6868, Standard specification for labeling of end items that incorporate plastics and polymers as coatings or additives with paper and other substrates designed to be aerobically composted in municipal or industrial facilities, ASTM International, West Conshohocken, PA.

[88] F.V. Ferreira, L.S. Cividanes, R.F. Gouveia, L.M.F. Lona, An overview on properties and applications of Poly(butylene adipate-co-terephthalate)–PBAT based composites, Polym. Eng. Sci. 59(2019) 7-15. https://doi.org/10.1002/pen.24770

[89] ISO 20200, Plastics -- Determination of the degree of disintegration of plastic materials under simulated composting conditions in a laboratory-scale test.

[90] E.L. Sánchez-Safont, J. González-Ausejo, J. Gámez-Pérez, J.M. Lagarón, L. Cabedo, Poly(3-Hydroxybutyrate-co-3-Hydroxyvalerate)/purified cellulose fiber composites by melt blending: characterization and degradation in composting conditions, J. Renew. Mater. 4 (2016) 123-132. https://doi.org/10.7569/JRM.2015.634127

[91] E. Lidón Sánchez-Safont, A. Arrillaga, J. Anakabe, J. Gamez-Perez, L. Cabedo, PHBV/TPU/cellulose compounds for compostable injection molded parts with improved thermal and mechanical performance, J. Appl. Polym. Sci. 136 (2019) 47257-47269. https://doi.org/10.1002/app.47257

[92] E.L. Sánchez-Safont, A. Arrillaga, J. Anakabe, L. Cabedo, J. Gamez-Perez, Toughness enhancement of PHBV/TPU/cellulose compounds with reactive additives for compostable injected parts in industrial applications, Int. J. Mol. Sci. 19 (2018) 2102-2121. https://doi.org/10.3390/ijms19072102

[93] C.S. Wu, Renewable resource-based composites of recycled natural fibers and maleatedpolylactide bioplastic: Characterization and biodegradability, Polym. Degrad. Stab. 94 (2009) 1076-1084. https://doi.org/10.1016/j.polymdegradstab.2009.04.002

[94] C.S. Wu, Characterization and biodegradability of polyester bioplastic-based green renewable composites from agricultural residues, Polym. Degrad. Stab. 97 (2012) 64-71.

[95] C.S. Wu, Preparation, characterization and biodegradability of crosslinked tea plant-fibre-reinforced polyhydroxyalkanoate composites, Polym. Degrad. Stab. 98 (2013) 1473-1480. https://doi.org/10.1016/j.polymdegradstab.2013.04.013

[96] C.S. Wu, H.T. Liao, The mechanical properties, biocompatibility and biodegradability of chestnut shell fibre and polyhydroxyalkanoate composites, Polym. Degrad. Stab. 99 (2014) 274-282. https://doi.org/10.1016/j.polymdegradstab.2013.10.019

[97] C.S. Wu, Renewable resource-based green composites of surface-treated spent coffee grounds and polylactide: Characterisation and biodegradability, Polym. Degrad. Stab. 121 (2015) 51-59. https://doi.org/10.1016/j.polymdegradstab.2015.08.011

[98] C.S. Wu, H.T. Liao, Y.X. Cai, Characterisation, biodegradability and application of palm fibre reinforced polyhydroxyalkanoate composites, Polym. Degrad. Stab. 140(2017) 55-63. https://doi.org/10.1016/j.polymdegradstab.2017.04.016

[99] L. Joyyi, M.Z.A. Thirmizir, M.S. Salim, L. Han, P. Murugan, K. Kasuya, F.H.J. Maurer, M.I.Z. Arifin, K. Sudesh, Composite properties and biodegradation of biologically recovered P(3HB-co-3HHx) reinforced with short kenaffibers, Polym. Degrad. Stab. 137 (2017) 100-108. https://doi.org/10.1016/j.polymdegradstab.2017.01.004

[100] M.N. Prabhakar, J. Song, Fabrication and characterisation of starch/chitosan/flax fabric green flame-retardant composites, Int. J. Biol. Macromol. 119 (2018) 1335-1343. https://doi.org/10.1016/j.ijbiomac.2018.07.006

[101] T. Bayerl, M. Geith, A.A. Somashekar, D. Bhattacharyy, Influence of fibre architecture on the biodegradability of FLAX/PLA composites, Int. Biodeterior. Biodegradation 96 (2014) 18-25 https://doi.org/10.1016/j.ibiod.2014.08.005

[102] Z.N. Terzopoulou, G.Z. Papageorgiou, E. Papadopoulou, E. Athanassiadou, E. Alexopoulou, D.N. Bikiaris, Green composites prepared from aliphatic polyesters and bastfibers, Ind. Crops Prod. 68 (2015) 60-79. https://doi.org/10.1016/j.indcrop.2014.08.034

[103] M. Oliveira, C. Mota, A.S. Abreu, J.M. Nobrega, Development of a green material for horticulture, J. Polym. Eng. 35 (2014) 1-6. https://doi.org/10.1515/polyeng-2014-0262

[104] J.S. Won, J.E. Lee, D.Y. Jin, S.G. Lee, Mechanical properties and biodegradability of the kenaf/soy protein isolate-PVA biocomposites, Int. J. Polym. Sci. 2015 (2015) 1-11. https://doi.org/10.1155/2015/860617

[105] D. Hammiche, A. Boukerrou, B. Azzeddine, N. Guermazi, T. Budtova, Characterization of polylactic acid green composites and its biodegradation in a bacterial environment, Int. J. Polym. Anal. Ch. 24 (2019) 236-244. https://doi.org/10.1080/1023666X.2019.1567083

[106] M. Akonda, S. Alimuzzaman, D.U. Shah, A.N.M. Masudur Rahman, Physico-mechanical, thermal and biodegradation performance of random flax/polylactic acid and unidirectional flax/polylactic acid biocomposites, Fibers. 6 (2018) 98-116. https://doi.org/10.3390/fib6040098

[107] H. Ibrahim, S. Mehanny, L. Darwish, M. Farag, A comparative study on the mechanical and biodegradation characteristics of starch-based composites reinforced with different lignocellulosicfibers, J. Polym. Environ. 26 (2018) 2434-2447. https://doi.org/10.1007/s10924-017-1143-x

[108] J.R.N. de Macedo, D.J. dos Santos, D. dos Santos Rosa, Poly(lactic acid)–thermoplastic starch–cotton composites: Starch-compatibilizing effects and composite biodegradability, J. Appl. Polym. Sci. 136 (2019) 47490-47499. https://doi.org/10.1002/app.47490

[109] H.M. Nakhoda, Y. Dahman, Mechanical properties and biodegradability of porous polyurethanes reinforced with green nanofibers for applications in tissue engineering, Polym. Bull. 73 (2016) 2039-2055. https://doi.org/10.1007/s00289-015-1592-0

[110] Z.N. Terzopoulou, G.Z. Papageorgiou, E. Papadopoulou, E. Athanassiadou, M. Reinders, D.N. Bikiaris, Development and study of fully biodegradable composite materials based on poly(butylene succinate) and hemp fibers or hemp shives, Polym. Compos. 37 (2016) 407-421. https://doi.org/10.1002/pc.23194

[111] L. Xie, H. Xu, Z.P. Wang, X. J. Li, J.B. Chen, Z.J. Zhang, H.M. Yin, G.J. Zhong, J. Lei, Z.M. Li, Toward faster degradation for natural fiber reinforced poly(lactic acid) biocomposites by enhancing the hydrolysis-induced surface erosion, J. Polym. Res. 21(2014) 357-371. https://doi.org/10.1007/s10965-014-0357-z

[112] D.K. Debeli, Z. Qin, J. Guo, Study on the pre-treatment, physical and chemical properties of ramie fibers reinforced poly (lactic acid) (PLA) biocomposite, J. Nat. Fibers. 15 (2018) 596-610. https://doi.org/10.1080/15440478.2017.1349711

[113] H.L. Boudjema, H. Bendaikha, Composite materials derived from biodegradable starch polymer and atriplexhalimusfibers, e-Polymers. 15 (2015) 419-426. https://doi.org/10.1515/epoly-2015-0118

[114] M. Seggiani, P. Cinelli, N. Mallegni, E. Balestri, M. Puccini, S. Vitolo, C. Lardicci, A. Lazzeri, New bio-composites based on polyhydroxyalkanoates and Posidoniaoceanica fibres for applications in a marine environment, Materials. 10 (2017) 326-338. https://doi.org/10.3390/ma10040326

[115] H.A. Kratsch, J.A. Schrader, K.G. McCabe, G. Srinivasan, D. Grewell, W.R. Graves, Performance and biodegradation in soil of novel horticulture containers made from bioplastics and biocomposites, Horttechnology. 15 (2015) 119-131. https://doi.org/10.21273/HORTTECH.25.1.119

[116] C.S. Wu, Preparation and characterizations of polycaprolactone/green coconut fiber composites, J. Appl. Polym. Sci. 115 (2010) 948-956. https://doi.org/10.1002/app.30955

[117] C.S. Wu, Preparation and characterization of polyhydroxyalkanoate bioplastic-based green renewable composites from rice husk, J. Polym. Environ. 22 (2014) 384-392. https://doi.org/10.1007/s10924-014-0662-y

[118] C.S. Wu, Aliphatic polyester-based green renewable eco-composites from agricultural residues: Characterization and assessment of mechanical properties, J. Polym. Environ. 21(2013) 421-430. https://doi.org/10.1007/s10924-012-0515-5

[119] C.S. Wu, Enhanced interfacial adhesion and characterisation of recycled natural fibre-filled biodegradable green composites, J. Polym. Environ. 26 (2018) 2676-2685. https://doi.org/10.1007/s10924-017-1160-9

[120] C.S. Wu, Solar energy tube processing of lemon residues for use as fillers in polyester based green composites, Polym. Bull. 75 (2018) 5745-5761. https://doi.org/10.1007/s00289-018-2359-1

[121] M.N. Prabhakar, A.R. Shah, J. Song, Improved flame-retardant and tensile properties of thermoplastic starch/flax fabric green composites, Carbohydr. Polym. 168 (2017) 201-211. https://doi.org/10.1016/j.carbpol.2017.03.036

[122] M.N. Prabhakar, J. Song, Fabrication and characterisation of starch/chitosan/flax fabric green flame-retardant composites, Int. J. Biol. Macromol. 119 (2018) 1335-1343. https://doi.org/10.1016/j.ijbiomac.2018.07.006

[123] T. Bayerl, M. Geith, A.A. Somashekar, D. Bhattacharyy, Influence of fibre architecture on the biodegradability of FLAX/PLA composites, Int. Biodeterior. Biodegradation. 96 (2014) 18-25. https://doi.org/10.1016/j.ibiod.2014.08.005

[124] M.A Gunning, L.M Geever, J.A Killion, J.G Lyons, C.L Higginbotham, The effect of processing conditions for polylactic acid based fibre composites via twin-screw extrusion, J. Reinf. Plast. Compos. 33 (2014) 648-662. https://doi.org/10.1177/0731684413512225

[125] E. Abraham, P.A. Elbi, B. Deepa, P. Jyotishkukar, L.A. Pothen, S. Thomas, S.S. Narine, X-ray diffraction and biodegradation analysis of green nanocomposites of natural rubber/nanocellulose, Polym. Degrad. Stab. 97 (2012) 2378-2387. https://doi.org/10.1016/j.polymdegradstab.2012.07.028

[126] E. Abraham, M.S. Thomas, C. John, L.A. Pothen, O. Shoseyov, S. Thomas, Green nanocomposites of natural rubber/nanocellulose: Membrane transport, rheological and thermal degradation characterisations, Ind. Crops Prod. 51 (2013) 415-424. https://doi.org/10.1016/j.indcrop.2013.09.022

[127] Z. Yue-Hong, Z. Wu-Quan, G. Zhen-Hua, G. Ji-You, Effects of crosslinking on the mechanical properties and biodegradability of soybean protein-based composites, J. Appl. Polym. Sci. 132 (2015) 41387-41395. https://doi.org/10.1002/app.41387

Advanced Applications of Bio-degradable Green Composites
Materials Research Foundations 68 (2020) 45-59

Materials Research Forum LLC
https://doi.org/10.21741/9781644900659-2

Chapter 2

Applications of Poly-3-Hydroxybutyrate Based Composite

Ranjna Sirohi[1], Jai Prakash Pandey[1], Ayon Tarafdar[2], Raveendran Sindhu[3]*,
Parameswaran Binod[3] and Ashok Pandey[4]

[1]Department of Post Harvest Process and Food Engineering, G.B. Pant University of Agriculture and Technology, Pantnagar, Uttarakhand-263 145, India

[2]Department of Food Engineering, National Institute of Food Technology Entrepreneurship and Management, Sonepat, Haryana-131 028, India

[3]Microbial Processes and Technology Division, CSIR-National Institute of Interdisciplinary Science and Technology (CSIR-NIIST), Trivandrum – 695 019, India

[4]Center for innovation and translational research, CSIR- Indian Institute of Toxicology Research (CSIR-IITR), 31 MG Marg, Lucknow-226 001, India

sindhurgcb@gmail.com; sindhufax@yahoo.co.in

Abstract

Increase in global population lead to the increase in the usage of non-biodegradable plastics and this excess usage lead to societal as well as environmental concerns. Several attempts are reported for the disposal of these via recycling, burning or land filling. None of these strategies provide an ideal solution. Hence, an alternative strategy is usage of biodegradable polymers. Poly-3-hydroxybutyrate represents bioplastics synthesized by bacteria which serves as energy and carbon storage compound which can be mobilized and used when there is a limitation of carbon. Microbes produce these polymers when supplemented with excess carbon source and limitations of nutrients like nitrogen or phosphorous. This chapter discusses the importance and applications of poly-3-hydroxybutyrate based composites in diverse fields like agriculture, food packaging, medicine, industry, nanotechnology, other miscellaneous applications as well as conclusion and future perspectives.

Keywords

Polyhydroxybutyrate, Biopolymer, Composite, Applications, Biodegradable

Contents

1.   Introduction

Polyhydroxyalkanoates or PHAs are naturally occurring intracellular bacterial polymers which are produced as storage polyesters utilizing broad range of microorganisms normally under unhinged growth conditions [1]. Poly-3-hydroxybutyrate (PHB), 3-hydroxyvalerate (3HV) and, poly (3-hydroxybutyrate-co-3- hydroxyvalerate) (PHB-co-V), are popular copolymers of the PHA family. These biopolymers show good biocompatibility and biodegradability. They resemble various material properties to conventional plastics that have attracted widespread attention and can be used in agricultural, industrial, and medical applications. Both PHB homopolymer and PHBV co-polyester have a gamut of applications due to their good combination of mechanical, thermal, biological and surface properties. Following are the important areas in which PHB can be used frequently: (a) PHB can be used as filler for non-biodegradable plastics; (b) disposable but degradable packages; (c) agricultural systems for extensive release of fertilizers (macro and micro nutrients) and agrochemicals; (d) medical appliances and drug delivery systems [2]. Polyhydroxybutyrate is well suited for industrial applications

Materials Research Forum LLC
https://doi.org/10.21741/9781644900659-2

due to its excellent gas barrier characteristics, high crystallinity (50-70%), and its polypropylene like physical attributes [3, 4, 5, 6, 7]. However, PHB is also associated with some disadvantages, such as high fragility, low thermal stability, brittleness and low nucleation density which are derived from its high purity and stereo-chemical regularity. Numerous methods have been proposed to remediate these disadvantages, including the synthesis of copolymers, chemical grafting and polymer blending. Among these, blending with other polymers is considered a cost-effective and simpler to use processing technology at an industrial scale to enhance the inherent properties of polyhydroxybutyrate.

Medical applications of microbial PHB/PHAs and its copolymers as biocompatible and biodegradable material have attracted a lot of attention in the last decade. A demanding combination of biodegradable and biomedical characteristics of PHB is an important tool in the design and development of new medical equipment. From previous years, PHB is being used extensively in the development of novel devices including sutures, slings, repair patches and devices, cardiovascular patches, cartilage repair devices, bone marrow scaffolds, orthopaedic pins, guided tissue repair/regeneration devices, adhesion barriers, stents, tendon repair devices, nerve guides, bone marrow scaffolds, and wound dressings [8]. In addition, PHB has the potential to controlled release of different drugs and encapsulation of drugs; therefore it can be extensively used for the conglomeration of therapeutic systems for sustained drug delivery [9].

Mechanical and physicochemical characteristics of PHBs render them suitable replacements for petrochemical based bulk plastics such as polypropylene and polyethylene. In the near future, environmental friendliness will be one of the leading factors which are estimated from human endeavours. In this aspect, several efforts are made towards the research and development of biopolymers with suitable properties or popularly known as "green" polymers with dissimilarity to the petro-chemically derived conventional polymers [10, 11].

2. Importance of Poly-3-hydroxybutyrate based composites

PHB is stiff and brittle due to its high crystallinity. This diminishes the span of PHB applications. Substantially, PHB has lower resistance to thermal degradation which creates most critical problem in processing of PHB. Due to the delicate innate properties of PHB, it is usually blended with fibers, chitosan, lignin, plasticizers and other additives (starch, cellulose) to enhance its mechanical and physical properties and, to reduce the processing temperature. The blends or composites are produced using solvent casting. Polyethylene glycol (PEG) shows good compatibility with polyhydroxybutyrate [12]. Araujo et al. [13] prepared a biodegradable poly (3-hydroxybutyrate) composite with 8 to 55 % chemically polymerized hydrochloric acid-doped polyaniline nanofibers of size 70–100 nm. Rodrigues et al. [14] reported good miscibility of PHB with low content of polyethylene glycol (2-5 %). Ikejima et al. [15] developed biodegradable polyester/polysaccharide composites and films using microbial produced polyhydroxybutyrate with chitin and chitosan. Godbole et al. [16] reported that PHB blending cost may be reduced using 70:30 ratio of PHB-starch is maintained. Wei et al. [17] recovered low-value waste byproduct by potato peel fiber fermentation and augmented them with PHB to form durable biocomposites. Although the developed biocomposites showed poor mechanical characteristics; extremely high biodegradation rate was seen at fiber content >15 % as compared with pure PHB.

3.   Applications of Poly-3-hydroxybutyrate based composites

Polyhydroxybutyrate has wide spectrum of applications including its use in agricultural, automotive interior, electrical appliances, construction materials, containers, sanitary items and novel packaging materials, among others. PHB is the first commercially accessible bio-based biodegradable product which has desired thermo-mechanical properties for a plethora of applications including medical materials, disposable items, film based products (e.g., shopping bags, covering films, and compost bags), and packaging materials [6]. PHB can be molded, extruded, made into thin films and spun into fibers. Microbial produced PHB also exhibits compatibility with body tissues, which facilitates its use in medical areas such as surgical sutures, controlled drug released systems, wound dressings and possible ocular devices [18].

3.1   Applications in agriculture

PHB can be used for the manufacture of many agricultural components such as mulch film, plant pots, seed trays, ties, clips and also be cost savers and biodegradable without harming the agricultural field [19]. Currently, materials used in agricultural fields are picked up manually after the completion of harvesting or disposed off altogether with the green waste at higher cost while PHB based composites (bioplastics) can be disposed off into normal compost along with plant waste with no additional for their disposal [20]. In Spain, starch based biodegradable plastics have been sold for several years; is easily degradable at desired rate and also shows high yields [21]. Recently, Procter and Gamble have developed Nodax[TM], which can be utilized to produce biodegradable films. Nodax[TM] is a copolymer, which mostly contains poly-3-hydroxybutyrate and minor amounts of MCL monomers. Nodax[TM] degrades anaerobically and therefore can serve as a coating material for fertilizers (urea), herbicides and insecticides which are commonly used in rice fields. A specialized application of poly-3-hydroxybutyrate and its copolymer poly-3-hydroxyvalerate in the agricultural sector is the controlled release of insecticides which can be administered while sowing of the crops. PHB/PHAs can also be used in the form of bacterial inoculants to intensify nitrogen fixation in plants [7]. For agricultural purposes, the bacterial culture used in inoculant preparation should be able to resist stressful environments.

3.2   Applications in food packaging

PHB could be a good packaging material for the food grade products and many researchers found that it is stiffer and much less flexible in comparison to conventional plastics [22, 23, 24]. Due to its high cost and easily broken quality bound the use of PHB for food packaging industries [25]. However, it is predictable that the PHB production will increase four times by 2021 as compare to 2016 [26] and, therefore, it is expected that the cost of PHB/PHAs will reduce. PHB films made from polylactic acid and PHB are successfully used as a packaging material for the sterilization of meat salad [27]. Bucci et al. [28] reported that PHB films can be used in place of polypropylene for the packaging of fat based products like mayonnaise, cheese, etc. The authors observed the change in physical, mechanical and sensory properties throughout their study. Khosravi et al. [29] stated that nanoparticles could improve the overall properties of PHB for food packaging. Haugaard et al. [30] reported that the PHB materials are suitable for liquid acidic and fatty food stuff. They used PHB for the packaging of orange juice and dressing which showed the same stability as in HDPE. Muizniece et al. [31] also reported that PHB materials can be used to store dairy products. Fabra et al. [32] introduced the polyhydroxyalkoanate based multilayer structures where the authors used zein fibre interlayer for increasing the oxygen barrier.

3.3   Applications in medicine

There are many applications of PHB in healthcare. It can be used for the development of implanted medical devices for dental, orthopedic, cranio-maxillo-facial, skin surgery and hernioplastic [7]. Some potential medical devices have been developed using PHB as base material: biodegradable plates and screws for bone and cartilage fixation, bioresorbable surgical sutures [33, 34], biodegradable membranes for periodontal treatment, PHB coated surgical meshes for hernioplastic surgery [9] and wound coverings [35].

In most of the studies, composites of PHB and hydroxyapatite (HA) are used as scaffolds to treat bone abnormalities [36]. Copolymer of PHB and polyglycolic acid (PGA) can produce pulmonary valve leaflets and pulmonary artery scaffolds in sheep [37]. Freier et al. [38] reported that PHB films provide a platform to patch a large bowel defect in rats and can be easily degraded *in vivo*. Tepha a company in the USA manufactures numerous medical devices using PHA/PHB and blending of both with other copolymers. Tepha FLEX® surture is the first well-known product, and approved by the United State Food and Drug Administration (FDA), made-up from P (4HB). Other surgical meshes and films are also fabricated by Tepha, using PHA/PHB.

PHB based controlled drug releasing devices are rapidly used for the human health care. "Drug release" refers to the process by which the drug solutes transfer from an initial location in the polymer matrix to the outer surface of the polymer system and subsequently, to the medium [39]. This process is influenced by various parameters such as the release environment, structural characteristics of the system, and probable interactions between them. Michalak et al. [40] prepared synthetic controlled drug release model PHB-amine conjugate which contain hydrolysable imine bond and reported that the unique properties of PHB carrier would be powerful tool for the development of novel drug conjugates. Bonartsev et al. [5] designed the biodegradable microspheres using poly (3-hydroxybutyrate) (PHB) as a base material for the controlled release of antithrombotic drug (dipyridamole). They reported that there will be possibility to elaborate the novel injectable therapeutic system for a local, long-term, anti-proliferative action by using the combination of biodegradable PHB and dipyridamole (DPD). Lins et al. [41] prepared composites using microparticles of poly (3-hydroxybutyrate) (PHB) in which ketoprofen (KET) drug was coated with cross linked layer of chitosan for the application of controlled drug-release devices. They found that the release of drug from the PHB based composite which was coated with cross linked layer of chitosan was slow and sustainable, which shows high potential for drug delivery. Gangrade and Price used PHA based microspheres as carriers for steroids [42]. Recently, Lu et al. reported PHA as an anticancer drug delivery device [43]. They developed a PHA nanoparticle based continued drug release system of P13K inhibitor (TGX221) and which could obstruct the proliferation of cancer cell lines. Use of PHB based thermogels as delivery agents of chemotherapeutics for the effective shrinkage of tumours was reported by Wu et al. [44]. This thermogel biopolymer possesses *in vitro* biocompatibility with very low cytotoxicity in HEK293 cells. The developed thermogel is also able manifest controllable release of doxorubicin (DOX) and paclitaxel (PTX) by adjusting the polymer concentrations. An *in vivo* application of this PHB based thermogel has been performed in rodents in which it has shown significant reduction in the rate of the tumour growth.

3.4   Applications in industry

Nowadays, biocomposites are very high in demand. There are so many industries that produce composites for their different purposes. As a Bioenvelop company (Canada) produces BioP articles for making the food containers, National starch company (UK) using P (3HB) based composites for making the packaging materials, Plastobag industries (India) making plastic bags (plasto-bag) using poly (3-hydroxybutyrate). Biomer, a German company produces poly (3-hydroxybutyrate) (3HB) from the strain *Alcaligenes latusan* which can accumulate nearly 90 % P(3HB) in the cell. The developed polymer is useful in manufacture of combs, pens and bullets [7]. Metabolix, is another such company in the US that produces metabolix PHA, a composites of P (3HB) and poly 3-hydroxyoctanoate) using recombinant *E. coli* K12. The product is an elastomer which is FDA approved for the production of food additives. Tsinghua University (China), Guangdon Jiangmen center for Biotech Development (China), KIAST (Korea) and Pocter and Gamble (USA) collaboratively prepared P (3HB- 3HH$_x$) composite which is used to make flushable, binders, non wovens, flexible packaging, thermoformed articles, medical devices and synthetic paper. Due to the piezoelectric nature of the developed composites, they can be used for the production of various instruments/items such as instruments for acceleration measurement, keyboard pressure sensors, testing of materials, microphone, headphones, loudspeakers, shock wave sensors, gas lighters, ultrasonic detectors, oscillators, for atomization of liquids and ultrasonic therapy.

3.5   Applications in nanotechnology

Nanotechnology deals with the science of matter manipulation at the nanoscale (1-1000 nm). Nanoparticles of PHB and its co-polymers have found numerous applications in medical science including but not limited to wound management, vascular system, efficient drug delivery, tissue repair and engineering, orthopedic products and ultrasound imaging. The unique properties of PHB such as biodegradability, uniform chirality and biocompatibility enable its use for different nanotechnology based applications such as the development of nanospheres for efficient drug delivery [45]. The core material of drugs can be made using PHB which slowly degrades in the body thereby releasing the inner constituents uniformly and without local irritation [46]. Since PHB molecules are hydrophobic in nature, its intestinal uptake has been found to be much more efficient than hydrophilic particles [47]. Immobilization of proteins using nano-beads has also been reported as a significant application of PHB. Polyhydroxybutyrate based endotoxin free smart beads produced from gram positive bacteria are used in biomedical applications for administering drug and vaccines removing the limitations of conventional therapeutic treatments.

PHB can also be used for manufacture of nanocomposite materials. PHB based green nanocomposites are a new generation of environment friendly materials and thus have wide range of applications. In contrast to general polyolefins, PHBs adhere better to lignocellulosic fibers which make them suitable for use in bionanocomposite materials. The high cyrstallinity and brittle nature of PHB could pose difficulties in the production of composites but these limitations can be gradually resolved by the addition of macromolecules and nanoparticles into the composite architecture. Such incorporations could enhance the ductility, melt viscosity and thermal stability of the resulting material which could further broaden the scope of PHB based materials. In this regard, PHB based nanocomposite scaffolds are used in repair of spinal cord injuries and bone related applications due to their lower roughness and lower agglomeration which improves cell attachment [48]. Nanofiber scaffolds of PHB show good viability of natural stem cells which grow on this material like natural extracellular matrix [49]. Nanocomposites from PHB also show improved physical properties with increased the elongation rate of approximately 5-80 % as compared with PHB used individually.

3.6   Miscellaneous applications

3.6.1   Aquaculture

The anti-adhesive property of PHB is used in aquaculture sector to prevent biofilm development. In contrast to conventional aquacultural procedures which focuses on pathogenic growth inhibition, PHB is used for deterring biofilm development thereby, hindering biomass formation. Lee et al. reported that the surface accumulation of methyl groups in PHB is responsible for its anti-adhesive nature [50]. PHB consists of short chain fatty acids which show strong inhibition to pathogenic *V. campbellii* [51]. PHB has also been reported to have anti-adhesive impact on pathogens encountered in shrimp aquaculture. Moreover, PHB has shown potential as an anti-microbial agent for Nile tilapia larviculture where it has improved the survival rate of feeding challenged tilapia larvae by 20 % over control.

3.6.2   Antifouling

Fouling in pipelines, industrial equipment and shipments, processing and desalination plants causes serious economic losses. Antifouling technology was developed to remediate this problem and was established in association with maritime transportation of goods as well as people [52]. PHB based nanocomposite coatings are effective antifoulants and are used on marine vessel hulls to ward off marine organisms. Toxicants including copper and tributylin present in the traditional paint matrix of these coatings gradually leach the biocide to prevent settlement of microorganisms [53]. PHB antifouling nanocomposites can also be prepared from marine microbes to delineate the use of metal coupons which can make the composites environmentally less harmful.

3.6.3 Tissue engineering

PHBs are hydrophobic, brittle and have a slow degradation rate which hinders its direct application in biomedical procedures. The use of PHB in addition to various polymers such as alginates, cellulose, chitosan and pectin have been studied and have found to generate promising results in the domain of tissue engineering. For instance, electrospun pectin-PHB nanofibres have demonstrated potential to be used as a scaffold in tissue engineering for the repair of damaged retinal cells [54]. PHB and its copolymers are also used for the development of sutures, cardiovascular patches, orthopaedic pins, bone plating system and surgical mesh, among others. PLA/PHBV blends show efficient mechanism in maintaining cell growth and production of extracellular matrix on the cells. Cheng et al. reported that PHB matrix cultured lung fibroblast cells of Chinese Hamster exhibits excellent cell compatibility with incorporation of PEG [55]. PHBV films have also shown to be a promising temporary substrate for the transplantation of retinal cells. It has been established that materials produced on PHB matrix can produce desirable results of bone tissue adaption with no chronic inflammation after implantation [56].

Conclusions and future perspectives

Though PHB based composites finds wide range of applications in different sectors, product cost is the main challenge limiting commercialization. Several R and D activities are going on to reduce the cost. Another limitation is the production of different monomers by microbes, rather than a single type of PHB/PHA. Costs of carbon source in the medium contribute to 60 % of the overall production cost. Utilization of waste carbon sources could be an alternative strategy to address this issue. Another solution is improvisation of existing technologies to develop cost-effective down-stream strategies, since the down-stream processing costs around 20% of the overall production cost. Development of more efficient bioreactors, developing hyper-producing strains and polymers for diversion application are other viable alternative strategies for cost-reduction.

Acknowledgement

Ranjna Sirohi thanks CSIR for providing fellowship under direct SRF scheme. Raveendran Sindhu acknowledges DST for sanctioning a project under DST WOS-B scheme.

References

[1]   L.L. Madison, G. W. Huisman, Metabolic engineering of poly (3-hydroxyalkanoates): from DNA to plastic, Microbiol. Mol. Biol. Rev. 63 (1999) 21-53. https://doi.org/10.1128/MMBR.63.1.21-53.1999

[2]   K. Sudesh, H. Abe, Y. Doi, Synthesis, structure and properties of polyhydroxyalkanoates: biological polyesters, Prog. Polym. Sci. 25 (2000) 1503. https://doi.org/10.1016/S0079-6700(00)00035-6

[3]   H. Tsuji, Y. Ikada, Crystallization from the melt of poly (lactide)s with different optical purities and their blends, Macromol. Chem. Phys. 197 (1996) 3483. https://doi.org/10.1002/macp.1996.021971033

[4]   T. Freier, C. Kunze, C. Nischan, S. Kramer, K. Sternberg, M. Sass, U.T. Hopt, K.P. Schmitz, In vitro and in vivo degradation studies for development of a biodegradable patch based on poly (3-hydroxybutyrate), Biomaterial. 23 (2002) 2649-57. https://doi.org/10.1016/S0142-9612(01)00405-7

[5]   A.P. Bonartsev, V.L. Myshkina, D.A. Nikolaeva, E.K. Furina, T.A. Makhina, V.A. Livshits, A.P. Boskhomdzhiev, E.A. Ivanov, A.L. Iordanskii, G.A. Bonartseva, Biosynthesis, biodegradation, and application of poly(3-hydroxybutyrate) and its copolymers- natural polyesters produced by diazotrophic bacteria, Communicating Current Research and Educational Topics and Trends in Applied Microbiology A. Mendez-Vilas (Ed.) (2007).

[6]   P.R. Patnaik, "Intelligent" descriptions of microbial kinetics in finitely dispersed bioreactors: neural and cybernetic models for PHB biosynthesis by *Ralstonia eutropha*, Microb. Cell Fact. 6 (2007) 23-25. https://doi.org/10.1186/1475-2859-6-23

[7]   Y. Chaudhari, B. Pathak, M.H. Fulekar, PHA- Production, application and its bioremediation in environment, I. Res. J. Environment Sci. 1 (2012) 46-52.

[8]   G.Q. Chen, Q. Wu, The application of polyhydroxyalkanoates as tissue engineering materials, Biomaterials. 26 (2005) 6565-6578. https://doi.org/10.1016/j.biomaterials.2005.04.036

[9]   A.P. Bonartsev, G.A. Bonartseva, T.K. Makhina, V.L. Mashkina, E.S. Luchinina, V.A. Livshits, A.P. Boskhomdzhiev, V.S. Markin, A.L. Iordanskii, New poly-(3-hydroxybutyrate)-based systems for controlled release of dipyridamole and indomethacin, Prikl. Biokhim. Mikrobiol. 42 (2006) 710-715. https://doi.org/10.1134/S0003683806060159

[10] A.L. Iordanskii , Impact of structure and morphology upon water transport in polymers with moderate hydrophilicity. From traditional to novel environmentally friendly polymers, in: Water Transport in Synthetic Polymers. A.L. Iordanskii, O.V. Startsev, G.E. Zaikov (Eds.). Nova Science Publishers Inc., New York, 2004, pp. 1-13.

[11] A.L. Iordanskii, A.N. Schegolikhin, A.P. Bonartsev, G.A. Bonartseva, V.L. Myshkina, T.K. Makhina, S.P. Novikova. Proceedings of the 4th Moscow international congress "Biotechnology: state of the art and prospects of development". Moscow, Russia, 12-16 March 2007, part 2, p. 248.

[12] L. Zhang, X. Deng, S. Zhao, Z. Huang, Biodegradable polymer blends of poly(3-hydroxybutyrate) and poly(DL-lactide*)*-co-poly (ethylene glycol), J. Appl. Polym. Sci. 65 (1997) 1849–1856. https://doi.org/10.1002/(SICI)1097-4628(19970906)65:10<1849::AID-APP1>3.0.CO;2-F

[13] P.L.B. Araujo, C.R.P.C. Ferreira, E.S. Araujo. Biodegradable conductive composites of poly(3-hydroxybutyrate) and polyaniline nanofibers: Preparation, characterization and radiolytic effects, Express Polym. Lett. 5 (2011) 12–22. https://doi.org/10.3144/expresspolymlett.2011.3

[14] J. Rodrigues, D. Parra, A. Lugao, Crystallization on films of PHB/PEG blends evaluation by DSC, J. Therm. Anal. Calorim. 79 (2005) 379–381. https://doi.org/10.1007/s10973-005-0069-z

[15] T. Ikejima, K. Yagi, Y. Inoue, Thermal properties and crystallization behavior of poly (3-hydroxybutyric acid) in blends with chitin and chitosan, Macromol. Chem. Phys. 200 (1999) 413-421. https://doi.org/10.1002/(SICI)1521-3935(19990201)200:2<413::AID-MACP413>3.0.CO;2-Q

[16] S. Godbole, S. Gote, M. Latkar, T. Chakrabarti, Preparation and characterization of biodegradable poly-3-hydroxybutyrate starch blend films, Bioresour. Technol. 86 (2003) 33-37. https://doi.org/10.1016/S0960-8524(02)00110-4

[17] L. Wei, S. Liang, A.G. McDonald. Thermo physical properties and biodegradation behavior of green composites made from polyhydroxybutyrate and potato peel waste fermentation residue, Ind. Crops. Prod. 69 (2015) 91-103. https://doi.org/10.1016/j.indcrop.2015.02.011

[18] J. Lee, H. Lim, J. Hong. Application of non-singular transformation to on-line optimal control of poly-hydroxybutyrate fermentation, J. Biotechnol. 55 (1997) 135–150. https://doi.org/10.1016/S0168-1656(97)00064-3

[19] S.J. Yaradoddi, S. Hugar, N.R. Banapurmath, A.M. Hunashyal, M.B Sulochana, A.S. Shettar, S.V. Ganachari, Alternative and Renewable Bio-based and Biodegradable Plastics. Publisher: Springer International Publishing AG L.M.T. Martinez et al. (eds.), Handbook of Ecomaterials, 2018. https://doi.org/10.1007/978-3-319-48281-1_150-1

[20] Bioplastics, Brochure of the FNR. https://mediathek.fnr.de/media/downloadable/files/samples/b/r/brosch.biokunststoffe-web-v01_1.pdf, 2005 (accessed 23 August 2019).

[21] S. Kasirajan, M. Ngouajio, Polyethylene and biodegradable mulches for agricultural applications: a review, Agron. Sustain. Dev. 32 (2012) 501–529. https://doi.org/10.1007/s13593-011-0068-3

[22] D. Kamravamanesh, P. Stefan, K. Tamas, D. Irina, K. Paul, M. Lackner, C. Herwig, Increased poly-beta-hydroxybutyrate production from CO_2 in randomly mutated cells of cyanobacterial strain Synechocystis sp. PCC 6714: mutant generation and characterization, Bioresour. Technol. 266 (2018) 34-44. https://doi.org/10.1016/j.biortech.2018.06.057

[23] D.Z. Bucci, L.B.B. Tavares, I. Sell, Biodegradation and physical evaluation of PHB packaging, Polym. Test. 26 (2007) 908-915. https://doi.org/10.1016/j.polymertesting.2007.06.013

[24] F. Masood, Polyhydroxyalkanoates in the food packaging industry, Nanotechnology Applications in Food. (2007)153-177. https://doi.org/10.1016/B978-0-12-811942-6.00008-X

[25] P.S.D.O. Patricio, F.V. Pereira, M.C. dos Santos, P.P. de Souza, J.P.B. Roa, R.L. Orefice, Increasing the elongation at break of polyhydroxybutyrate biopolymer: Effect of cellulose nanowhiskers on mechanical and thermal properties, J. Appl. Polym. Sci. 127 (2013) 3613–3621. https://doi.org/10.1002/app.37811

[26] European Bioplastics. Available online: http://www.european-bioplastics.org/market/ (accessed on June 2017).

[27] V. Levkane, B.S. Muizniece, L. Dukalska, Pasteurization effect to quality of salad with meat and mayonnaise, Foodbalt. (2008) 69-73.

[28] D.Z. Bucci, L.B.B. Tavares, I. Sell, PHB packaging for the storage of food products, Polym. Test. 24 (2005) 564–571. https://doi.org/10.1016/j.polymertesting.2005.02.008

[29] K. Khosravi-Darani, D.Z. Bucci, Application of poly(hydroxyalkanoate) in food packaging: Improvements by nanotechnology, Chem. Biochem. Eng. 29 (2015) 275–285. https://doi.org/10.15255/CABEQ.2014.2260

[30] K. Haugaard, B. Danielsen, G. Bertelsen, Impact of polylactate and poly(hydroxybutyrate) on food quality, Eur. Food Res. Technol. 216 (2003) 233-240. https://doi.org/10.1007/s00217-002-0651-6

[31] B. Muizniece, L. Dukalska, Impact of biodegradable PHB packaging composite materials on dairy product quality, Proceedings of the Latvia University of Agriculture. 16 (2006) 79-87.

[32] M.J. Fabra, A. Lopez-Rubio, J.M. Lagaron, Nanostructured interlayers of zein to improve the barrier properties of high barrier polyhydroxyalkanoates and other polyesters, J. Food Eng. 127 (2014)1–9. https://doi.org/10.1016/j.jfoodeng.2013.11.022

[33] L. Galbraikh, M. Fedorov, G. Vikhoreva, N. Kildeeva, A. Maslikova, G. Bonartseva, Modeling of surface modification of sewing thread, Fibre Chem. 37 (2005) 441. https://doi.org/10.1007/s10692-006-0017-0

[34] A.V. Rebrov, V.A. Dubinsky, Y.P. Nekrasov, G.A. Bonartseva, M. Stamm, E.M. Antipov, *Vysokomol. Soedin.* 44 (2002) 347-351 (in Russian).

[35] N.R. Kil'deeva, G.A. Vikhoreva, L.S. Gal'braikh, A.V. Mironov, G.A. Bonartseva, P.A. Perminov, A.N. Romashova. Preparation of porous films for use as wound coverings Romashova, Prikl. Biokhim. Mikrobiol. 42 (2006) 716. https://doi.org/10.1134/S0003683806060160

[36] S.K. Misra, S.P. Valappil, I. Roy, A.R. Boccaccini. Polyhydroxyalkanoate (PHA)/inorganic phase composites for tissue engineering applications, Biomacromolecules. 7 (2006) 2249-58. https://doi.org/10.1021/bm060317c

[37] R. Sodian, J.S. Sperling, D.P. Martin, U. Stock, J.E. Mayer, J.P. Vacanti. Tissue engineering of trileaflet heart valve early in vitro experiences with a combined polymer, Tissue Eng. 5 (1999) 489-94. https://doi.org/10.1089/ten.1999.5.489

[38] T. Freier, C. Kunze, C. Nischan C, S. Kramer, K. Sternberg, M. Sass, U.T. Hopt, K.P. Schmitz. In vitro and in vivo degradation studies for development of a biodegradable patch based on poly(3-hydroxybutyrate), Biomaterials. 23 (2002) 2649-57. https://doi.org/10.1016/S0142-9612(01)00405-7

[39] K.W. Fu Y, Drug release kinetics and transport mechanisms of nondegradable and degradable polymeric delivery systems, Polym. Test. 7 (2010) 429-444. https://doi.org/10.1517/17425241003602259

[40] M. Michalak, A.A. Marek, J. Zawadiak, M. Kawalec, P. Kurcok. Synthesis of PHB-based carrier for drug delivery systems with pH-controlled release, Eur. Polym. J. 49 (2013) 4149–4156. https://doi.org/10.1016/j.eurpolymj.2013.09.021

[41] L.C. Lins, G.C. Bazzo, P.L.M. Barreto, A.T.N. Pires. Composite PHB/Chitosan microparticles obtained by spray drying: effect of chitosan concentration and crosslinking agents on drug relesase, J. Braz. Chem. Soc. 25 (2014) 1462-1471. https://doi.org/10.5935/0103-5053.20140129

[42] N. Gangrade, J.C. Price, Poly (hydroxybutyrate-hydroxyvalerate) microspheres containing progesterone: preparation, morphology and release properties, J. Microencapsul. 8 (1991) 185–202. https://doi.org/10.3109/02652049109071487

[43] X.Y. Lu, E. Ciraolo, R. Stefenia, G.Q. Chen, Y. Zhang, E. Hirsch, "Sustained release of PI3K inhibitor from PHA nanoparticles and *in vitro* growth inhibition of cancer cell lines, Appl. Microbiol. Biotechnol. 89 (2011) 1423-1433.

[44] Y.L. Wu, H. Wang, Y.K. Qiu, S. S. Liow, Z. Li, X. J. Loh, PHB-based gels as delivery agents of chemotherapeutics for the effective shrinkage of tumors, Adv. Healthcare Mater. 5 (2016) 2679–2685. https://doi.org/10.1002/adhm.201600723

[45] Z.A. Raza, S. Abid, I.M. Banat. Polyhydroxyalkanoates: Characteristics, production, recent development and applications, Int. Bioterior. Biodegrad. 126 (2018) 45-56. https://doi.org/10.1016/j.ibiod.2017.10.001

[46] A. des Rieux, V. Fievez, M. Garinot, Y.J. Schneider, V. Preat. Nanoparticles as potential oral delivery systems of proteins and vaccines: A mechanistic approach, J. Control. Release. 116 (2006) 1–27. https://doi.org/10.1016/j.jconrel.2006.08.013

[47] M. Zohri. Polymeric NanoParticles: Production, applications and advantage, Internet. J. Nanotechnol. 3 (2009) 1-14.

[48] S.K. Srivastava, A.D. Tripathi. Effect of saturated and unsaturated fatty acid supplementation on bio-plastic production under submerged fermentation, Biotech. 3 (2013) 389–397. https://doi.org/10.1007/s13205-012-0110-4

[49] X.Y. Xu, X.T. Li, S.W. Peng, J.F. Xiao, C. Liu, G. Fang, Chen, K.C. Chen, The behaviour of neural stem cells on polyhydroxyalkanoate nanofiber scaffolds, Biomaterials. 31 (2010) 3967–3975. https://doi.org/10.1016/j.biomaterials.2010.01.132

[50] C. Lee, B. Song, J. Jegal, Y. Kimura, Cell adhesion and surface chemistry of biodegradable aliphatic polyesters: discovery of particularly low cell adhesion behavior on poly (3-[RS]-hydroxybutyrate), Macromol. Res. 21 (2013) 305–1313. https://doi.org/10.1007/s13233-013-1181-8

[51] T. Defoirdt, D. Halet, H. Vervaeren, N. Boon, T.V. deWiele, P. Sorgeloos, P. Bossier,W. Verstraete, The bacterial storage compound poly-β-hydroxybutyrate protects *Artemia franciscana* from pathogenic Vibrio campbellii, Environ. Microbiol. **9** (2007) 445–452. https://doi.org/10.1111/j.1462-2920.2006.01161.x

[52] G. Kavitha, R. Rengasamy, D. Inbakandan. Polyhydroxybutyrate production from marine source and its application, Int. J. Biol. Macromol. 111 (2018) 102-108. https://doi.org/10.1016/j.ijbiomac.2017.12.155

[53] J.A. Lewis, Marine biofouling and its prevention on underwater surfaces, Mater. Forum. 22 (1998) 41–61.

[54] S.Y. Chan, B.Q. Y. Chan, Z. Liu, B.H. Parikh, K. Zhang, Q. Lin, X. Su, D. Kai, W.S. Choo, D.J. Young, X.J. Loh. Electrospun pectin-polyhydroxybutyrate nanofibers for retinal tissue engineering, ACS Omega. 2 (2017) 8959-8968. https://doi.org/10.1021/acsomega.7b01604

[55] G.X. Cheng, Z.J. Cai, L. Wang, Biocompatibility and biodegradability of poly (hydroxybutyrate)/poly(ethylene glycol) films, J. Mater. Sci. Mater. Med. 14 (2003) 1073–1078. https://doi.org/10.1023/B:JMSM.0000004004.37103.f4

[56] C. Doyle, E.T. Tanner, W. Bonfield, *In vitro* and *in vivo* evaluation of polyhydroxybutyrate and of polyhydroxybutyrate reinforce with hydroxyapatite, Biomat. 12 (1991) 841–847. https://doi.org/10.1016/0142-9612(91)90072-I

Advanced Applications of Bio-degradable Green Composites　　　　　Materials Research Forum LLC
Materials Research Foundations **68** (2020) 60-84　　　　　https://doi.org/10.21741/9781644900659-3

Chapter 3

Plant Fibre Based Biodegradable Green Composites

M. Harikrishna Kumar[1], C. Moganapriya[1], R. Rajasekar[1*], T. Mohanraj[2]

[1]Department of Mechanical Engineering, Kongu Engineering College, Erode, TamilNadu, India

[2] Department of Mechanical Engineering, Amrita School of Engineering, Amrita Vishwa Vidyapeetham, Amritanagar Campus, Coimbatore, Tamil Nadu, India

*rajasekar.cr@gmail.com

Abstract

Bio-composite are fabricated by reinforcing natural fiber into a biopolymer matrix. Bio-composite are manufactured using mechanical and melt mixing technique followed by hot pressing at elevated temperature. Fiber volume fraction is varied in the biopolymer matrix to find the optimum technical properties. Chemically treated natural fiber used as a reinforcing material in preparing bio-composite results in higher technical properties when compared to untreated fiber bio-composite. As weight percentage of fiber increases physic-mechanical property enhances. Physic-mechanical property reduces at particular weight percentage of natural fiber. The trend is observed due to the agglomeration of fiber in the bio-polymer matrix. The Bio-degradation study also proves that bio-composite prepared using natural fiber and biopolymer can be used as an alternative material for conventional thermoplastic composite.

Keywords

Natural Fiber, Biopolymer, Bio-Composite, Biodegradable

Contents

1.   Introduction

Fiber Reinforced Polymer (FRP) composites are produced by reinforcing fiber into a polymer matrix which are used in number of applications such as aircraft, automobile, structural, etc. FRP composite possess high strength and stiffness combined with low density when compared to conventional materials which results in weight reduction of a finished component [1-6]. Owing to advantages, FRP composites have become more popular among the industrialist and researchers. Over 250 million tons of commodity plastics are produced worldwide. For most of us, life without polymers and plastics is unimaginable. These polymers are long lasting and show high persistence in the environment, which is seen as an advantage in many applications such as pipes, aircraft, etc. However, when these plastics are disposed into the environment, it will accumulate in nature for decades. Since these materials are not readily biodegradable due to its low rate of degradability. It creates several issues to the living bodies [7-11]. Since FRP composite contains two different components, it is difficult to reuse and recycle and is directly dumped as waste plastics [12]. The disposal of plastics and adverse effect on environment has created a major challenge among researchers. It leads to focus their research and development on bio-degradable composite for various applications [12-14]. Evolution of biodegradable polymer composites has started and is still growing in the market. Even though biodegradable materials have their unique technical properties, due to thier high cost biodegradable materials are unable to commercialize in the market [13, 15-19]. Bio-composites are derived from renewable resources instead of petroleum resources. This provides benefits to the industry as well as to the environment. Thes bio-composites are used in automobiles such as dashboards, door panels, etc. Automobile industry has started to replace glass fiber with natural plant fibers such as jute, flax, hemp, sisal, etc. But still if bio-composites become cost effective they can penetrate into the market as a replacement to conventional materials [14].

2.   Degradation mechanism of biopolymer

Degradation of polymer happens by thermal an activation process, such as, hydrolysis method, biological activity, oxidation, photolysis or radiolysis method. Mechanism of degradation of polymer can also be referred to as environmental degradation which is due to the combined effect of biotic and non-biotic processes [20]. Degradation of biopolymer is encountered in four steps (i) Bio-deterioration (ii) De-polymerization (iii) Bio-assimilation and (iv) Mineralization is shown in Fig. 1. Bio-deterioration leads to formation of microbial film which leads to degradation of polymeric material which is divided into smaller particles. Microbial film produces enzymes which act as catalyst for de-polymerization process of polymer chain into oligomers, dimers or monomers. In bio-assimilation step uptake of small molecules into microbial film which in turn produces primary and secondary metabolites. Metabolites are mineralized to form CO_2, CH_4, H_2O, and N_2 which in turn are released to the environment [8].

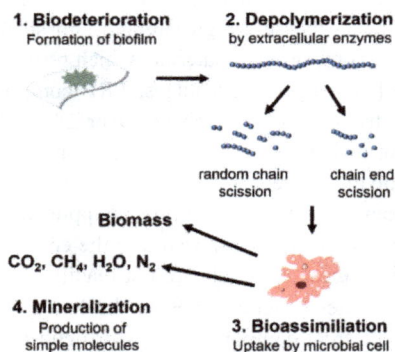

1. Biodeterioration
Formation of biofilm

2. Depolymerization
by extracellular enzymes

random chain
scission

chain end
scission

Biomass

CO_2, CH_4, H_2O, N_2

4. Mineralization
Production of
simple molecules

3. Bioassimiliation
Uptake by microbial cell

Figure 1. Degradation mechanism of Biopolymer [8]

3.   Fabrication procedure

Fabrication methodology of bio-composite is shown in Fig. 2. Bio-polymer material was cut according to requirement of the die size. Natural fibers consist of hemicellulose, lignin and pectin content which affects the bonding between matrix and material which in turn reduces the physic-mechanical properties of the composite. In order to enhance the physic-mechanical properties of the composite generally, fibers are alkaline treated which in turn partially removes the hemicellulose, lignin and pectin from the surface of the fiber. Treated and non-treated fibers are cut into different sizes. Bio-polymer and short natural fibers are placed in the mold layer after layer uniformly. Mold together with biopolymer and natural fiber is placed onto the heating unit and heated up to 110°C to 200°C under a pressure of 1.3MPa for 10 minutes. Temperature and pressure range is selected based on the type of biopolymer and fiber used. Then the mold is removed from the heating unit and the pressure is maintained until the setup is cooled to room temperature. Ruihua Hu has fabricated hemp fiber reinforced polylactic acid composites by using this methodology [21].

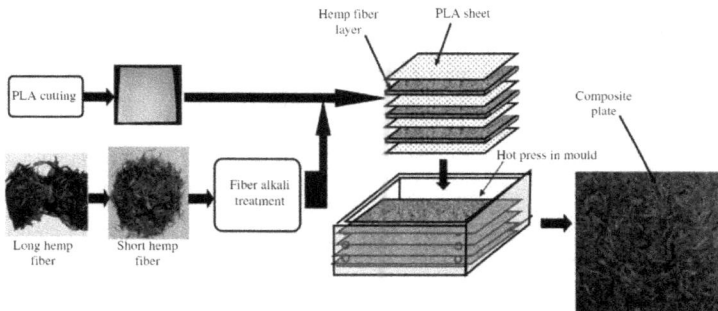

Figure 2. Fabrication Procedure [21]

4. Biocomposite

A composite material can be defined as a combination of two or more materials that results in better properties than those of the individual components used alone. The main constituents of a composite are matrix and reinforcement material are shown in Fig. 3. Matrix materials play an important role while the load is applied to the material. It should possess the ability of deformation under applied load and transfer the load to the fibers to distribute stress concentration [1]. Matrix materials can be derived from petroleum products or from natural resources. Non-biodegradable and biodegradable polymers are extracted from petroleum products. Polymers like polypropylene (PP), polyethylene, Poly(tetrafluoro ethylene), etc., belongs to non-biodegradable polymers. Polymers like poly lactic acid (PLA), Polyesters, Polycaprolactone (PCL), etc., belongs to biodegradable polymers. The reinforcing phase provides the strength and stiffness. In most cases, the reinforcement is harder, stronger and stiffer than the matrix. The reinforcement material is usually a fiber. Fibers used in a bio-composite is usually extracted from plants, animal and geological process [1, 12-14].

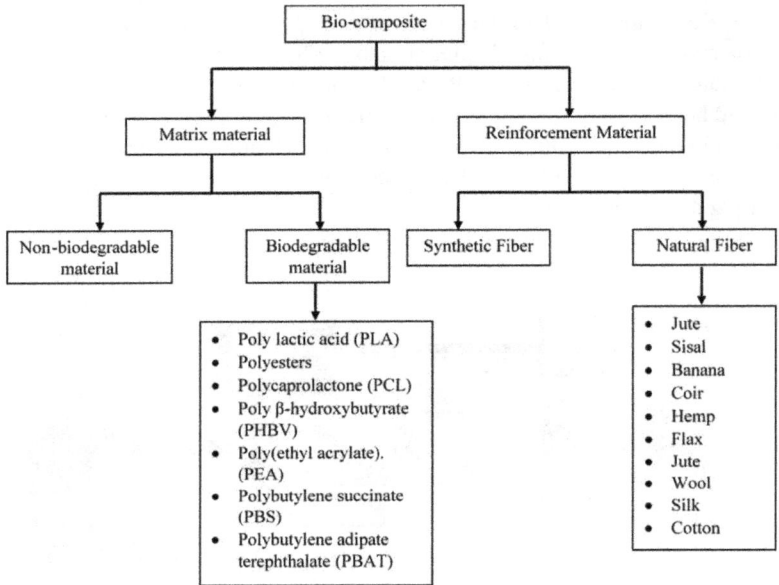

Figure 3. Constituents of Bio-composite

4.1   Poly lactic acid (PLA) reinfirced plant fiber biocomposite

Ruihua Hu has studied hemp fiber reinforced in PLA matrix. The hot press method was used to fabricate hemp fiber reinforced composites. Composites were prepared by using alkaline treated and untreated hemp fibers which were cut into 5-15 mm in length. Composites were also prepared by using different fiber volume fraction such as 30%, 40% and 50%. Alkaline treated fiber composite shows better tensile strength when compared to untreated fiber composite [22]. Composite with 40% fiber volume fabrication with alkaline treatment shows high tensile strength, elastic modulus and flexural strength when compared to 30% and 50% fiber volume fraction composite [21-23]. Fig. 4 shows the stress-strain curves for different volume fraction.

Shinji has prepared fiber reinforced composite by kneaf as a fiber and PLA as a matrix. Thermal studies were made on kneaf fiber and it was found that at 180°C tensile strength starts decreasing. Based on thermal studies of fiber, Shinji has decided the molding temperature as 160°C. Molding pressure was maintained at 10 MPa for 10 minutes. Kneaf/PLA composites were prepared by 30%, 50% and 70% fiber volume fraction and the fiber directions were maintained unidirectional. Bio-composite prepared with 70%

fiber volume fraction shows high strength in terms of tensile and flexural. Shinji has also examined the biodegradability of kneaf/PLA composite by using garbage-processing machine. Tensile strength and weight of the composite were found to be decreased by 91% and 38% after four weeks of composting by which Shinji has proven experimentally its biodegradability [24].

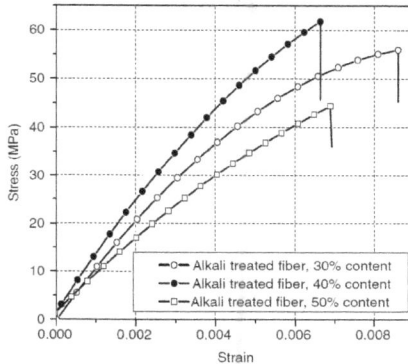

Figure 4. Stress-Strain Curve with different hemp fiber volume fraction [21]

Another work with natural fiber and PLA as a matrix has been reported by Oksman. Flax fiber is used as a reinforcement material for preparing composite. Additives like triacetin, glycerol triacetate ester was also used for preparing composite to reduce the brittleness of the PLA polymer. Long treated fibers were connected by using hand roving technique which is shown in Fig. 5. Oksman has employed twin-screw extruder and compression molding to manufacture bio-composite. Bio-composite were prepared by using 30% and 40% fiber volume fraction. Technical properties of Flax/PLA composite were compared with polypropylene (PP) / flax composite, since PP fiber composite were most commonly used in automobile parts. Study reveals that PLA/flax composite are 50% higher in strength when compared to PP/flax composite. PLA/flax composite prepared with 30% fiber volume shows increase in stiffness from 3.4 to 8.4 GPa. Impact strength of PLA/flax composite has not increased due to the usage of triacetin as an additive in PLA matrix. Gel permeation chromatography (GPC) analysis was carried out for the PLA/flax composite to study the biodegradability. Studies shows that PLA/flax composite was not degraded due to the preparation method. Method of preparation of composite plays a major role while preparing bio-composite. Fig. 6, shows the SEM image of tensile fractured surface of PLA/flax composite [2].

Figure 5. Roven Flax Fiber [2]

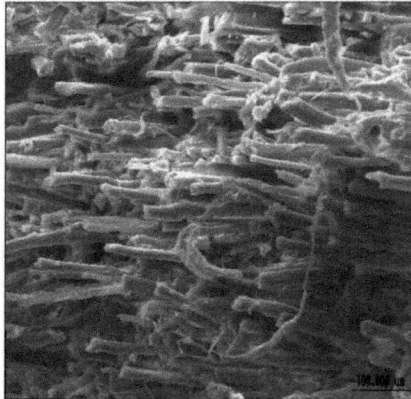

Figure 6. Fractured surface of PLA/flax composites[2]

David has combined jute fiber with l-polylactide to produce biodegradable composite materials. Bio-composite materials were produced by film stacking technique in which l-polylactide is converted into a film and jute mat is used as a reinforcement material as shown in Fig. 7. Film stacking techniques is followed by pre-compression, contact heating under vacuum, compression and cooling using press and removal of the composite from the press. These bio-composites were prepared with different process temperature such as 180°C, 190°C, 200°C and 210°C by maintaining 40% fiber volume fraction. Tensile strength and stiffness of PLA/jute composite is increased by 100% when compared to pure PLA. High tensile strength is observed when PLA/jute composite were manufactured at 210°C. Size exclusion chromatography analysis were carried out for PLA, jute mat and PLA/jute composite. David observed a decrease in molecular weight as the process temperature increases [25].

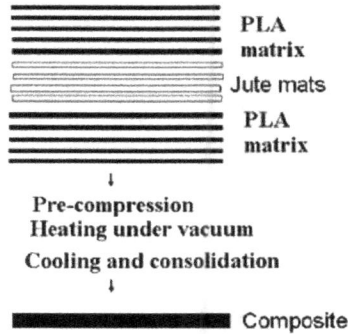

Figure7. Fabricating PLA/jute composites [25]

Bio-composite was also prepared by using banana fiber as reinforcement material in PLA matrix. In this PLA/banana composites were prepared by using alkali treated fibers. The banana fibers were treated with 4% NaoH solution for 45 minutes and washed in distilled water to reduce the pH value. The fibers were chopped to obtain an average length 1 cm. Polytetrafluoroethylene film is added as a coupling agent to alkaline treated fibers and kept in a sealed flask for about 12 hours. Coupled alkaline treated fibers were washed with acetone to remove compounds which are not covalently bonded and dried at 80°C. Then the coupled alkaline treated fibers are added into the PLA matrix along with dicumyl peroxide (DCP) by melt blended method for 15 minutes at 170°C in an internal mixer. Schematic representation of preparation is shown in Fig. 8. The bio-composites were prepared by maintaining weight ratio of fiber as 20, 40 and 60 phr based on the weight of the PLA. Thermal stability and mechanical properties of pure PLA is enhanced by reinforcement of the banana fiber in the PLA matrix. Mechanical strength and thermal stability of the bio-composite increase with increase in weight percentage of the fiber. Bio-composite prepared with 40 phr of fiber shows maximum tensile and flexural strength. But impact strength of the bio-composite starts decreasing with increase in weight percentage of fiber. Production cost of the bio-composite has drastically reduced due to the reinforcement of the banana fiber [26].

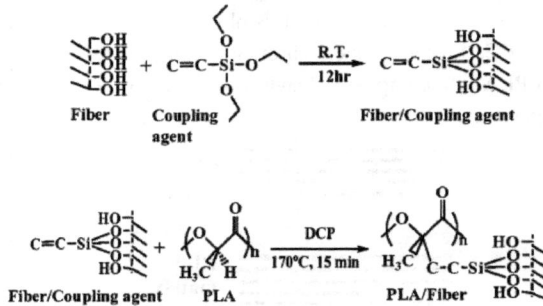

Figure 8. Reaction between fiber, coupling agent and PLA [26]

Yussuf has prepared bio composites using kneaf and rice husk as reinforcement material in the PLA matrix. Technical and biodegradability properties are compared. Fiber volume fraction was maintained as 20% between the PLA matrixes. Twin extruder and injection molding technique were used for preparing bio-composite. PLA/kneaf composite shows better mechanical properties when compared to PLA/rice husk composite. SEM images show that there is less interfacial bonding between fiber and PLA matrix which is shown in Fig. 9. PLA/kneaf and PLA/rice husk composite has less thermal stability when compared to pure PLA. PLA/rice husk composite shows high thermal degradation when compared to PLA/kneaf composite. Biodegradability studies were carried out for the keaf and rice husk composite by burrowing the composite in normal soil to stimulate natural biodegradation. Composite were recovered from the soil in different stages such as 10, 30 and 90 days. Studies reveal that reinforcement of kneaf or rice husk fibers in to PLA matrix improves the biodegradability when compared to pure PLA. PLA/kneaf composite shows better biodegradation rate when compared to PLA/rice husk composite. PLA/kneafcomposite shows better technical and biodegradation rate when compared to PLA/rice husk composite. It may be due to difference in chemical composition and low aspect ratio [27].

Nina Graupner has produced bio-composites using natural fibers and manmade cellulose fibers. PLA/natural fiber composite were manufactured by compression molding and PLA/manmade cellulose fiber composite were manufactured by injection molding technique. Bio-composites were manufactured using different fibers such as cotton, hemp, kneaf and manmade cellulose fiber (Lyocell). Fiber volume fraction is maintained as 40%. Various types of bio-composite were prepared by using fibers in different combination such as cotton 40% + PLA 60%, Lyocell 40% + PLA 60%, kenaf 40% + PLA 60%, hemp 40% + PLA 60%, hemp 20% + kenaf 20% + PLA 60% and hemp 20% + Lyocell 20% + PLA 60%. Descent improvement in tensile strength, young's modulus and impact strength is achieved for PLA/cotton composite when compared to pure PLA. High tensile strength, elongation at break and young's modulus is achieved for PLA/hemp, PLA/hemp + kneaf and PLA/kneaf composite when compared to PLA/cotton composite. High tensile strength, elongation at break and impact strength is observed for PLA/hemp + lyocell when compared to PLA/hemp. Property of fibers plays a major role in deciding the property of bio-composites. Table 1, depicts the different natural fiber composites, its volume fraction and strength which are most commercially used in automobile applications. Properties obtained from different bio-composite (PLA/cotton, PLA/lyocell, PLA/kneaf, PLA/hemp, PLA/kenaf + kneaf and PLA/hemp + lyocell) can be compared with commercially available natural fiber composites in order to find suitable applications [28].

Figure 9. SEM images of tensile fractured (a) PLA/kneaf (b) PLA/rice husk [27]

Table 1. Natural fiber composite and its properties which are commercially available [28]

Natural Fiber	Matrix	Volume Fraction (Fiber : Matrix)	Tensile Strength (N/mm^2)	Young's Modulus (N/mm^2)	Impact Strength (kJ/m^2)
Bast fibre	Polypropylene	30:70	18-25	--	8-25
Bast fibre	Polypropylene	50:50	2-30	--	25-35
Flax/hemp	Epoxy	65:35	40-60	--	14-20
Wood fibre	Acrylate	90:10	25-30	--	20-30
Wood + synthetic fibres	Acrylate	85:12	25-30	--	12-20
Hemp/kenaf	Acrylate	77:23	--	3000-1200	6-40

4.2   Poly-Butylene-Succinate (PBS) reinfirced plant fiber biocomposite

Bio-composites were manufactured using flax fiber as a reinforcement material in a PBS matrix. Alain has also fabricated bio-composites by varying the matrix material such as poly-propylene (PP) and PLA as matrix. Fiber volume fraction is approximately maintained as 25.5%. Bio-composites were fabricated using extrusion and injection method carried out at 140°C. Reinforcement of flax fiber in to PBS matrix enhances the mechanical properties. However, PBS possess low young's modulus, PBS/flax composite shows low rigidity and high strength, elongation at break and impact strength when compared PP/flax and PLA/flax composite. PBS is also mixed in PLA/flax composite in varying volume fraction. During blending PBS volume fraction is maintained equal to the fiber volume fraction. Blending of PBS/PLA/flax composite revealed positive results with increase in mechanical properties as shown in Fig. 10. Alain has concluded that blending of matrix enhances the properties of bio-composite [29].

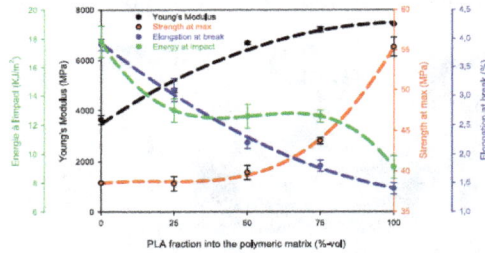

Figure 10. Technical properties of PBS/PLA/flax bio-composite [29]

Yanhong has reported the preparation and properties of PBS/sisal fiber reinforced bio-composite. Bio-composite were produced using two roll mixer and followed by hot pressing. PBS/sisal bio-composite were manufactured by maintaining 30% fiber volume fraction and by varying the two roll mixture temperature such as 170, 190, 200, 210, and 230°C for 6 minutes. Then the mixture were hot pressed at 160°C under 10 MPa for 5 minutes. The sisal fiber used for preparing the bio-composites were chopped to a length of 8 mm and are pretreated by exposing them to steam and dried in an oven at 100°C for 12 hours. Results revealed that mechanical properties of PBS/sisal fiber composites are improved as the mixing temperature of the two roll mixture increases. Interfacial bonding between PBS and sisal fiber is improved as the operating temperature increases. The mechanical properties of the PBS/sisal bio-composite increased for a mixing temperature of 200°C. Table 2, shows the results of mechanical properties of PBS/sisal fiber bio-composite fabricated under different mixing temperatures [30].

Table 2. Influence of temperature on PBS/sisal fiber bio-composite [30]

Mixing Temperature °C	Tensile Strength (Mpa)		Modulus (Mpa)		Elongation at Break %	Flexural Strength (Mpa)		Flexural Modulus (Mpa)		Impact Strength (kJ/m^2)	
	Neat PBS	PBS/ sisal	Neat PBS	PBS/ sisal	PBS/ sisal	Neat PBS	PBS/ sisal	Neat PBS	PBS/ sisal	Neat PBS	PBS/ sisal
170	39	34.6	297.7	958.4	5.5	17.5	40.6	558.6	1469	43.2	14.9
190	41.4	32.9	279.1	884.4	5.51	19.2	33.2	598.6	1459	39.9	16.3
200	38.5	47.8	293.5	973.2	8.23	18.8	43.7	643.9	1403	31.4	20.7
210	34.8	43.6	298.1	901.8	7.77	17.9	38.3	593.5	1420	28.8	12
230	25.7	42.4	314.3	858	7.47	18.7	41.3	605.3	1452	29.7	16

Feng Yan has explored morphology on rheological properties of sisal fiber reinforced on PBS bio-composite. Rheological behavior of PBS/sisal fiber bio-composite depends on fiber morphology. Fig. 11, shows the morphological comparison of (a) sisal fiber (b) steam exposed sisal fiber (c) baggase fiber and (d) steam exposed baggase fiber. From the images it is observed that steam exposed fibers are slender, flexible and easily deformed when compared to untreated fibers. Non-Newtonian index n value of fibers with high aspect ratio and large contact area with matrix is smaller. In general, non-newtonian index n value decreases as fiber volume fraction increases in base matrix. Instead non-newtonian index n value increases as fiber volume fraction is increases. It is due to the agglomeration of fibers in base matrix which reduces the contact surface area [31].

Kneaf fiber and PBS were dried under vacuum at 40°C for 48 hours before manufacturing bio-composite. Kneaf fiber reinforced in PBS matrix were manufacture by melt mixing technique followed by hot pressing under 10 MPa pressure at 150°C for 2 minutes. Zhichao has fabricated PBS/kneaf bio-composite by varying fiber volume fraction such as 10, 20 and 30%. PBS and kneaf fiber were dried under vacuum at 40°C during the manufacturing of bio-composite. PBS and kneaf fiber were fed in to single screw machine and hot pressed at 150°C for 2 minutes under 10 MPa. Tensile strength of 30% fiber volume fraction in PBS/kneaf composite are drastically increased when compared to pure PBS. From the SEM analysis it is observed that interfacial adhesion between kneaf fiber and PBS matrix needs improvement. This strategy is observed due to usage of untreated fiber as an reinforcement material [32].

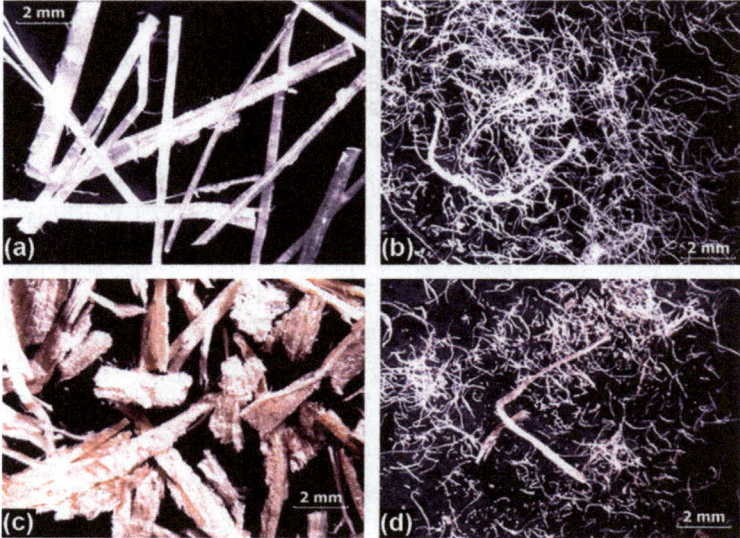

Figure 11. Morphological comparison of fibers [31]

Lifang Liu evaluated the biodegradability of bio-composite which is fabricated using PBS and jute fiber. Three diameter fibers 55, 48 and 40 µm fiber were used for fabrication of bio-composite. Fibers were treated with H_2SO_4 solution followed by soaking fibers in 5% NaOH and soaked in coupling agent KH-570. Biodegradability test were evaluated by burying the bio-composite in soil. Weight loss for the buried specimens is measured by different days such as 30, 60, 90, 120, 150 and 180 days as shown in Fig. 12. Among the prepared bio-composites, PBS/jute fiber composite prepared with 55 µm diameter shows higher weight loss when compared to bio-composites prepared by using 48 and 40 µm diameter. Less weight loss of bio-composite is observed as the fiber volume fraction increases. Maximum weight loss of bio-composite is observed for higher diameter and less fiber volume fraction. Weight loss is calculated by using the given formula [33].

Weight loss=(Weight Initial-Weight Final)/(Weight Final) X 100

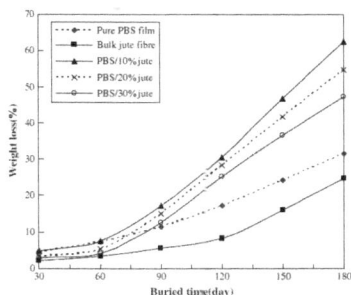

Figure 12. Weight loss of bio-composite [33]

PBS/coir fiber bio-composite were fabricated via hot pressing at 150°C for 10 minutes under 10 Mpa pressure. Coir fibers used in these studies were alkaline treated with 5% NaOH solution. Different category of soaking time were carried out for fibers say 24 hours, 48 hours, 72 hours and 96 hours. Bio-composites were also prepared by varying fiber volume fraction such as 10%, 15%, 20%, 25% and 30%. Among the NaOH treated fibers, 72 hours soaked coir treated fibers shows 55.6% higher interfacial shear strength when compared to untreated fibers as shown in Fig. 13. Technical properties of treated PBS/coir fiber bio-composite are higher than untreated PBS/coir fiber bio-composites. PBS/coir treated fiber with volume fraction 25% bio-composite shows superior mechanical properties when compared to 10%, 15%, 20% and 30% fiber volume fraction as shown in Fig. 14. Morphological studies also exhibits higher interfacial adhesion between fiber and matrix for treated fibers [34].

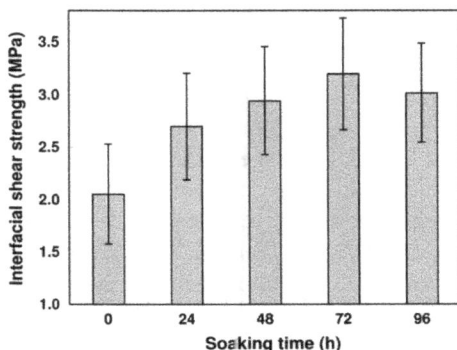

Figure 13. Interfacial strength of PBS/coir bio-composite[34]

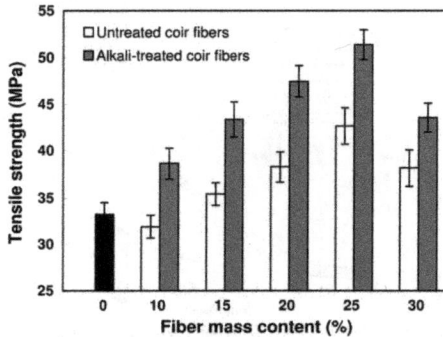

Figure 14. Tensile strength of PBS/coir bio-composite[34]

4.3   Polyhydroxybutyrate (PHB) reinfirced plant fiber biocomposite

Bio-composites were manufactured by using different natural fibers such as hemp and jute. Manmade cellulose (Lyocell) were also used as another fiber to prepare PHB bio-composite. Fiber volume fraction of hemp, jute and Lyocell are varied as 10%, 20% and 30% for the preparation of these bio-composites. Twin screw extrusion and injection molding technique were used for the fabrication process. Biodegradation studies were carried out using rotary aerated composter filled with organic waste. The rate of bio-degradation was measured once in a week. In all three cases modulus value increased as fiber content increases in the matrix. Among the three natural fibers, jute fiber shows high flexural modulus due to hydrogen bonding between fiber and matrix. Presence of natural fiber in PHB matrix increases the biodegradation rate. As fiber volume fraction in PHB matrix increases rate of biodegradability also increases which is shown in Fig. 15 [35].

Melo has investigated and fabricated PHB composite reinforcing carnauba fibers. Carnauba fibers were treated with concentration of 1%, 3% and 5% by alkali, acetylation and permanganate treatments. Property of carnauba fiber is compared with other natural fibers which are listed in Table 3. Fibers were chemically treated to improve the adhesion between fiber and the matrix. Bio-composite were prepared by hot pressing at 190°C for 2 minutes by maintaining fiber volume percentage as 10%. Thermal decomposition of carnauba fibers are found to be 240°C, which is very high when compared to all other natural fibers. Tensile strength of PHB/carnauba permanganate treatment fiber bio-composites was high when compared to untreated fibers and other treated fibers. SEM images of tensile fractured specimens proved that increase in adhesion between fiber and base matrix. Storage modulus of bio-composite are found to be higher at elevated temperatures when compared to pure PHB [36].

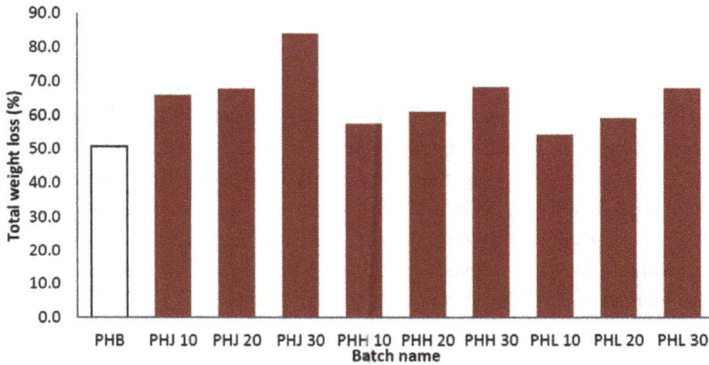

Figure 15. Weight loss percentage of bio-composite [35]

Table 3. Comparison of Natural Fiber Properties [36]

Natural Fiber	Fiber Density (g/cm^3)	Elongation (%)	Tensile Strength (MPa)	Young's Modulus (GPa)
Carnauba (FNT)	1.34	1.7–2.6	205–264	8.2–9.2
Carnauba (FTPH)	1.44	1.7–2.6	148–242	6.3–14.0
Cotton	1.5–1.6	3.0–10.0	287–597	5.5–12.6
Jute	1.3–1.46	1.5–1.8	393–800	10–30
Flaxa	1.4–1.5	1.2–3.2	345–1500	27.6–80
Hemp	1.48	1.6	550–900	70
Ramie	1.5	2.0–3.8	220–938	44–128
Sisal	1.33–1.5	2.0–14	400–700	9.0–38.0

Bio-composites were fabricated by reinforcing flax fiber in PHB base matrix. PHB is also called as biodegradable thermoplastic possessing similar physico-mechanical properties similar to that of non-biodegradable thermoplastics. Flax fiber possess low elastic property when compared to cotton fibers. Mechanical properties of the bio-composites were increases by incorporating flax fiber into the PHB matrix when compared to pure PHB. Elastic properties of fiber were also improved by synthesizing PHB [37].

4.4 Poly-Butylene Adipate Co-Terephatalate (PBAT) reinfirced plant fiber biocomposite

Agar polymers are well accepted by food and pharmaceutical industries. Agar polymers are synthesized by red seaweeds which belons to the genus of Gracilaria, Gelidium, and Pterocladia molecules. Bio-composites were manufactured by mixing of agar and PBAT using an extrusion and injection molding technique. Agar was mixed in PBAT in different weight percentage such as 10%, 20%, 30% and 40%. Mixing of agar into the PBAT matrix shows improvement in tensile strength and modulus. Stress strain relationship curve for PBAT/agar bio-composite and pure PBAT are shown in Fig. 16. Mechanical loss factor (tan δ) is reduced during the incorporation of agar particles into the PBAT matrix. Degradation temperature of PBAT/agar matrix lies in between of those individual values. Melting, crystallization temperature and viscosity of PBAT/agar bio-composite increases as weight percentage of agar increases [38].

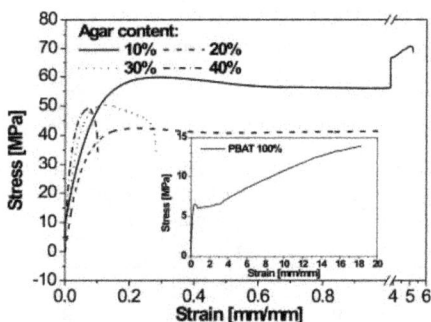

Figure 16. Stress strain curve for PBAT/agar and pure PBAT [38]

PLA/PBAT/Ramie bio-composites were prepared by Tao Yu. PLA/PBAT/ramie bio-composite are prepared by a two step process. First PLA/PBAT are mixed by twin extruder and then PLA/PBAT blend are mixed with ramie fiber using a two roll mill to prepare PLA/PBAT/ramie bio-composite as shown in Fig. 17. Blending of PBAT into the PLA matrix is varied in weight percentage such as 5%, 10% and 15%. Mechanical properties of the bio-composites are found to decreasing while blending PBAT into the PLA matrix. However toughness and thermal stability of the PLA/PBAT/Ramie bio-composite were improved. Adhesion between the ramie fiber and matrix has decreased due to the addition of PBAT in to PLA matrix which is observed in SEM images. It is also observed that glass transition temperature and the percentage crystallinity are found to decreases when compared to PLA/ramie fiber bio-composite [39].

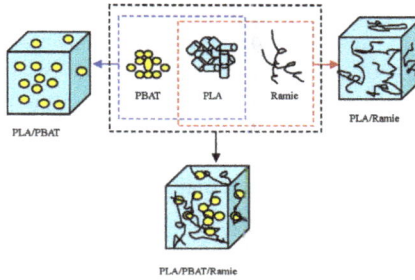

Figure 17. Manufacturing of PLA/PBAT/ramie bio-composite [39]

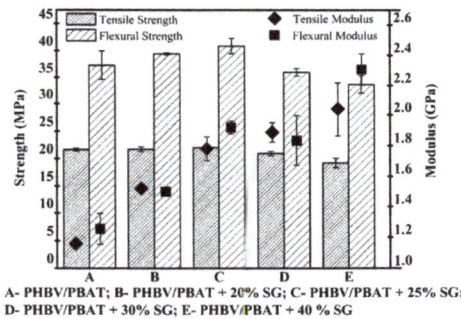

A- PHBV/PBAT; B- PHBV/PBAT + 20% SG; C- PHBV/PBAT + 25% SG;
D- PHBV/PBAT + 30% SG; E- PHBV/PBAT + 40 % SG

Figure 18. Tensile strength of PHBV/PBAT bio-composite [40]

Nagarajan has fabricated bio-composite using poly-3-hydroxybutyrate-co-3-hydroxyvalerate (PHBV) and PBAT matrix reinforced with switch grass fiber. Initailly two polymers PHBV and PBAT are blended. Fiber volume fraction was varied as 20%, 25%, 30%, 35% and 40% in PHBV/PBAT matrix. Poly diphenylmethane diisocyanate (pMDI) is used as a compatibilizer to solve the hydrophobic–hydrophilic disparity between fiber and matrix system. Switch grass fiber are added into the blended PHBV/PBAT matrix by melt mixing followed by melt mixing under controlled temperature maintained at 180°C. Then the blended fiber and composite were hot pressed at 45°C under 5 Mpa pressure. Technical properties of the PHBV/PBAT/pMDI/switch grass bio-composites were greatly enhanced when compared to PHBV/PBAT/switch grass bio-composites. Bio-composites with compatibilizer show greater improvement in properties. PHBV/PBAT/pMDI/switch grass compatabilized bio-composites prepared with 30% fiber volume fraction show high tensile strength and heat deflection temperature when compared to other bio-composites as shown in Fig. 18. Greater adhesion between fiber and polymer matrix is observed for the bio-composites prepared with compatibilizer [40].

4.5   Polycaprolactone (PCL) reinfirced plant fiber biocomposite

Thermoplastic starch (TPS) are blended with PCL to manufacture TPS/PCL bio-composite. Sisal fibers are reinforced into blended polymer TPS/PCL matrix. Sisal fibers are treated with NaOH solution and hydrogen peroxide to remove the hemicellulose content in the fiber. TPS/PCL/Sisal fiber bio-composites were fabricated by using twin extruder. Fiber volume fraction is varied in TPS/PCL matrix such as 5%, 10% and 20%. Bio-composites were fabricated by using treated fibers. Due to the reinforcement of treated fibers adhesion between the fiber and matrix is well enhanced. As fiber volume fraction increases, fiber got agglomerated in TPS/PCL matrix. Agglomeration of fibers in TPS/PCL matric resulted in decrease in technical properties. Bio-composites fabricated with 10% fiber volume fraction shows lower water absorption property. In CO_2 evolution study, bio-composite fabricated with TPS shows highest extent of degradation when compared to pure PCL. Due to its superior technical properties and water resistance properties TPS/PCL/Sisal fiber bio-composite can be used for variety of different engineering applications [41].

PLA and PCL polymer are blended and ramie fibers are reinforced into the blended matrix. PLA/PCL is blended by polymerization method. Ramie fibers are treated with coupling agents to improve the compatibility and interface between fiber and matrix. Ramie fibers are treated by dissolving KH550 in the mixed solvent (distilled water and ethanol) in the ratio of 1:9 to obtain a concentration of 1 weight percentage. Bio-composites were prepared by maintaining 15% fiber volume fraction with three length of fiber such as 1-2 mm, 5-6 mm and 10-12 mm length of fiber. Mechanical properties of PLA/PCL/ramie synthesized fiber are better than PLA/PCL/ramie un-synthesized fiber. Bio-composite of ramie synthesized fibers with length 5-6 mm in length shows superior properties when compared other bio-composite [42].

Conclusion

Number of research works on bio-composite has been carried out to find an alternative material for conventional composite material which is found to be hazardous to the environment due to its low rate of biodegradability. Bio-composites were prepared by using natural fiber and biodegradable polymer. Natural fiber are extracted from plants and trees such as sisal, jute, coir, flax, wool, hemp, banana, ramie, grass fiber, etc. Since these fibers are extracted from natural resources, they are biodegradable and do not spoil the environment as waste. The interest in natural fiber composites has started to grow in the late 1990's but originally the reinforcing material used was conventional thermoplastic which again partially pollutes the environment. Hence, researchers have focused on finding alternative material to conventional thermoplastic which results in evolution of PLA, PBS, PBAT, PCL, PHB biopolymers which are biodegradable when compared to polypropylene, polyvinyl chloride, etc. Work has been carried out by using natural fiber as a reinforcing material in biopolymer and number preparation techniques has been employed to improve the technical properties of such bio-composites. Our conclusion is based on studying previous literatures.

1. Bio-composite were prepared by mechanical and melt mixing followed by hot pressing at elevated temperature under constant pressure 10 MPa.

2. Natural fibers are reinforced in biopolymer matrix with various fiber volume fractions. Optimum weight percentage varies depending on the type of natural fiber and biopolymer used. As weight percentage of fiber increases the technical properties of the bio-composites increased. At particular loading technical properties starts decreasing and it is due to the agglomeration of fiber in the biopolymer matrix.

3. Natural fiber used in preparing bio-composite should undergo chemical treatment which results in enhancement in adhesion between fiber and matrix. Chemical treatment removes the impurities from the fiber surface and thus smoothen the fiber surface which results in better interfacial bonding.

4. Mechanical properties and water absorption studies were also found to be high while incorporating chemically treated fiber in biopolymers. Few were reported that technical properties of bio-composite is also compared with conventional thermoplastic composites and they found that bio-composites can be used as alternative material to conventional thermoplastic composite.

Biodegradability test were also carried out for the bio-composites and the rate of bio-degradability was found to be higher when compared to conventional thermoplastic composite.

References

[1] P.K. Bajpai, I. Singh, J. Madaan, Development and characterization of PLA-based green composites: A review, J. Thermoplast. Compos. Mater. 27 (2014) 52-81. https://doi.org/10.1177/0892705712439571

[2] K. Oksman, M. Skrifvars, J.F. Selin, Natural fibres as reinforcement in polylactic acid (PLA) composites, Compos Sci Tech. 63 (2003) 1317-1324. https://doi.org/10.1016/S0266-3538(03)00103-9

[3] A. Mohanty, M.A. Khan, G. Hinrichsen, Surface modification of jute and its influence on performance of biodegradable jute-fabric/Biopol composites, Compos Sci Tech. 60 (2000) 1115-1124. https://doi.org/10.1016/S0266-3538(00)00012-9

[4] S. Mukhopadhyay, R. Fangueiro, Physical modification of natural fibers and thermoplastic films for composites—a review, J. Thermoplast. Compos. Mater. 22 (2009) 135-162. https://doi.org/10.1177/0892705708091860

[5] M. Tajvidi, A. Takemura, Thermal degradation of natural fiber-reinforced polypropylene composites, J. Thermoplast. Compos. Mater.23 (2010) 281-298. https://doi.org/10.1177/0892705709347063

[6] R. Rahman, M. Hasan, M. Huque, N. Islam, Physico-mechanical properties of maleic acid post treated jute fiber reinforced polypropylene composites, J. Thermoplast. Compos. Mater. 22 (2009) 365-381. https://doi.org/10.1177/0892705709100664

[7] M. Cocca, F. De Falco, G. Gentile, R. Avolio, M.E. Errico, E. Di Pace, M. Avella, Degradation of Biodegradable Plastic Buried in Sand, Proceedings of the International Conference on Microplastic Pollution in the Mediterranean Sea, Springer, 2018, pp. 205-209. https://doi.org/10.1007/978-3-319-71279-6_28

[8] T.P. Haider, C. Völker, J. Kramm, K. Landfester, F.R. Wurm, Plastics of the future? The impact of biodegradable polymers on the environment and on society, Angew Chem Int Ed. 58 (2019) 50-62. https://doi.org/10.1002/anie.201805766

[9] A.L. Andrady, M.A. Neal, Applications and societal benefits of plastics, Phil Trans Biol Sci. 364 (2009) 1977-1984. https://doi.org/10.1098/rstb.2008.0304

[10] R.C. Thompson, Y. Olsen, R.P. Mitchell, A. Davis, S.J. Rowland, A.W. John, D. McGonigle, A.E. Russell, Lost at sea: where is all the plastic, Science. 304 (2004) 838-838. https://doi.org/10.1126/science.1094559

[11] J.R. Jambeck, R. Geyer, C. Wilcox, T.R. Siegler, M. Perryman, A. Andrady, R. Narayan, K.L. Law, Plastic waste inputs from land into the ocean, Science. 347 (2015) 768-771. https://doi.org/10.1126/science.1260352

[12] F. La Mantia, M. Morreale, Green composites: A brief review, Compos. Appl. Sci. Manuf. 42 (2011) 579-588. https://doi.org/10.1016/j.compositesa.2011.01.017

[13] G. Bogoeva-Gaceva, M. Avella, M. Malinconico, A. Buzarovska, A. Grozdanov, G. Gentile, M. Errico, Natural fiber eco-composites, Polymer compos. 28 (2007) 98-107. https://doi.org/10.1002/pc.20270

[14] G. Koronis, A. Silva, M. Fontul, Green composites: A review of adequate materials for automotive applications, Compos B Eng. 44 (2013) 120-127. https://doi.org/10.1016/j.compositesb.2012.07.004

[15] J. Mayer, Biodegradable materials: balancing degradability and performance, Trends Polym. Sci. 2 (1994) 227-235.

[16] H. Rozman, M. Lee, R. Kumar, A. Abusamah, Z.M. Ishak, The effect of chemical modification of rice husk with glycidyl methacrylate on the mechanical and physical properties of rice husk-polystyrene composites, J. Wood Chem. Tech. 20 (2000) 93-109. https://doi.org/10.1080/02773810009349626

[17] D.R. Carroll, R.B. Stone, A.M. Sirignano, R.M. Saindon, S.C. Gose, M.A. Friedman, Structural properties of recycled plastic/sawdust lumber decking planks, Resour Conservat Recycl. 31 (2001) 241-251. https://doi.org/10.1016/S0921-3449(00)00081-1

[18] A. Herrmann, J. Nickel, U. Riedel, Construction materials based upon biologically renewable resources—from components to finished parts, Polym Degrad Stabil. 59 (1998) 251-261. https://doi.org/10.1016/S0141-3910(97)00169-9

[19] C. Alves, A. Silva, L. Reis, M. Freitas, L. Rodrigues, D. Alves, Ecodesign of automotive components making use of natural jute fiber composites, J. Clean. Prod. 18 (2010) 313-327. https://doi.org/10.1016/j.jclepro.2009.10.022

[20] N. Nair, V. Sekhar, K. Nampoothiri, A. Pandey, Biodegradation of biopolymers, Current Developments in Biotechnology and Bioengineering, Elsevier2017, pp. 739-755. https://doi.org/10.1016/B978-0-444-63662-1.00032-4

[21] R. Hu, J.-K. Lim, Fabrication and mechanical properties of completely biodegradable hemp fiber reinforced polylactic acid composites, J Compos Mater. 41 (2007) 1655-1669. https://doi.org/10.1177/0021998306069878

[22] P. Wambua, J. Ivens, I. Verpoest, Natural fibres: can they replace glass in fibre reinforced plastics, Compos Sci Tech. 63 (2003) 1259-1264. https://doi.org/10.1016/S0266-3538(03)00096-4

[23] M.S. Huda, L.T. Drzal, A.K. Mohanty, M. Misra, Chopped glass and recycled newspaper as reinforcement fibers in injection molded poly (lactic acid)(PLA) composites: a comparative study, Compos Sci Tech. 66 (2006) 1813-1824. https://doi.org/10.1016/j.compscitech.2005.10.015

[24] S. Ochi, Mechanical properties of kenaf fibers and kenaf/PLA composites, Mech Mater. 40 (2008) 446-452. https://doi.org/10.1016/j.mechmat.2007.10.006

[25] D. Plackett, T.L. Andersen, W.B. Pedersen, L. Nielsen, Biodegradable composites based on L-polylactide and jute fibres, Compos Sci Tech. 63 (2003) 1287-1296. https://doi.org/10.1016/S0266-3538(03)00100-3

[26] Y.F. Shih, C.C. Huang, Polylactic acid (PLA)/banana fiber (BF) biodegradable green composites, J Polym Res. 18 (2011) 2335-2340. https://doi.org/10.1007/s10965-011-9646-y

[27] A. Yussuf, I. Massoumi, A. Hassan, Comparison of polylactic acid/kenaf and polylactic acid/rise husk composites: the influence of the natural fibers on the mechanical, thermal and biodegradability properties, J Polymer Environ. 18 (2010) 422-429. https://doi.org/10.1007/s10924-010-0185-0

[28] N. Graupner, A.S. Herrmann, J. Müssig, Natural and man-made cellulose fibre-reinforced poly (lactic acid)(PLA) composites: An overview about mechanical characteristics and application areas, Compos Appl Sci Manuf. 40 (2009) 810-821. https://doi.org/10.1016/j.compositesa.2009.04.003

[29] A. Bourmaud, Y.-M. Corre, C. Baley, Fully biodegradable composites: Use of poly-(butylene-succinate) as a matrix and to plasticize l-poly-(lactide)-flax blends, Ind. Crop. Prod. 64 (2015) 251-257. https://doi.org/10.1016/j.indcrop.2014.09.033

[30] Y. Feng, H. Shen, J. Qu, B. Liu, H. He, L. Han, Preparation and properties of PBS/sisal-fiber composites, Polymer Eng Sci. 51 (2011) 474-481. https://doi.org/10.1002/pen.21852

[31] Y.H. Feng, Y.J. Li, B.P. Xu, D.W. Zhang, J.P. Qu, H.Z. He, Effect of fiber morphology on rheological properties of plant fiber reinforced poly (butylene succinate) composites, Compos B Eng. 44 (2013) 193-199. https://doi.org/10.1016/j.compositesb.2012.05.051

[32] Z. Liang, P. Pan, B. Zhu, T. Dong, Y. Inoue, Mechanical and thermal properties of poly (butylene succinate)/plant fiber biodegradable composite, J Appl Polymer Sci. 115 (2010) 3559-3567. https://doi.org/10.1002/app.29848

[33] L. Liu, J. Yu, L. Cheng, X. Yang, Biodegradability of poly (butylene succinate)(PBS) composite reinforced with jute fibre, Polym Degrad Stabil. 94 (2009) 90-94. https://doi.org/10.1016/j.polymdegradstab.2008.10.013

[34] T.H. Nam, S. Ogihara, N.H. Tung, S. Kobayashi, Effect of alkali treatment on interfacial and mechanical properties of coir fiber reinforced poly (butylene succinate) biodegradable composites, Compos B Eng. 42 (2011) 1648-1656. https://doi.org/10.1016/j.compositesb.2011.04.001

[35] M.A. Gunning, L.M. Geever, J.A. Killion, J.G. Lyons, C.L. Higginbotham, Mechanical and biodegradation performance of short natural fibre polyhydroxybutyrate composites, Polym Test. 32 (2013) 1603-1611. https://doi.org/10.1016/j.polymertesting.2013.10.011

[36] J.D.D. Melo, L.F.M. Carvalho, A. Medeiros, C. Souto, C. Paskocimas, A biodegradable composite material based on polyhydroxybutyrate (PHB) and carnauba fibers, Compos B Eng. 43 (2012) 2827-2835. https://doi.org/10.1016/j.compositesb.2012.04.046

[37] M. Wróbel-Kwiatkowska, J. Zebrowski, M. Starzycki, J. Oszmiański, J. Szopa, Engineering of PHB synthesis causes improved elastic properties of flax fibers, Biotechnol Progr. 23 (2007) 269-277. https://doi.org/10.1021/bp0601948

[38] T. Madera-Santana, M. Misra, L. Drzal, D. Robledo, Y. Freile-Pelegrin, Preparation and characterization of biodegradable agar/poly (butylene adipate-co-terephatalate) composites, Polymer Eng Sci. 49 (2009) 1117-1126. https://doi.org/10.1002/pen.21389

[39] T. Yu, Y. Li, Influence of poly (butylenes adipate-co-terephthalate) on the properties of the biodegradable composites based on ramie/poly (lactic acid), Compos Appl Sci Manuf. 58 (2014) 24-29. https://doi.org/10.1016/j.compositesa.2013.11.013

[40] V. Nagarajan, M. Misra, A.K. Mohanty, New engineered biocomposites from poly (3-hydroxybutyrate-co-3-hydroxyvalerate)(PHBV)/poly (butylene adipate-co-

terephthalate)(PBAT) blends and switchgrass: Fabrication and performance evaluation, Ind Crop Prod. 42 (2013) 461-468. https://doi.org/10.1016/j.indcrop.2012.05.042

[41] A. de Campos, G.H. Tonoli, J.M. Marconcini, L.H. Mattoso, A. Klamczynski, K.S. Gregorski, D. Wood, T. Williams, B.S. Chiou, S.H. Imam, TPS/PCL composite reinforced with treated sisal fibers: Property, biodegradation and water-absorption, J Polymer Environ. 21 (2013) 1-7. https://doi.org/10.1007/s10924-012-0512-8

[42] H. Xu, L. Wang, C. Teng, M. Yu, Biodegradable composites: Ramie fibre reinforced PLLA-PCL composite prepared by in situ polymerization process, Polymer Bull. 61 (2008) 663-670. https://doi.org/10.1007/s00289-008-0986-7

Advanced Applications of Bio-degradable Green Composites   Materials Research Forum LLC
Materials Research Foundations **68** (2020) 85-103   https://doi.org/10.21741/9781644900659-4

Chapter 4

Biodegradable Materials for Planting Pots

B. Tomadoni[1*], D. Merino[1], C. Casalongué[2], V. Alvarez[1]

[1] Grupo de Materiales Compuestos Termoplásticos (CoMP), Instituto de Investigaciones en Ciencia y Tecnología de Materiales (INTEMA), Facultad de Ingeniería, Universidad Nacional de Mar del Plata (UNMdP) y Consejo Nacional de Investigaciones Científicas y Técnicas (CONICET), Colón 10850, (7600) Mar del Plata, Argentina

[2] Instituto de Investigaciones Biológicas. UE CONICET-UNMDP, Facultad de Ciencias Exactas y Naturales, Universidad Nacional de Mar del Plata, Deán Funes 3250, (7600) Mar del Plata, Argentina

bmtomadoni@fi.mdp.edu.ar

Abstract

The use of plastics based on non-renewable petroleum-sources in agriculture represents a growing threat to the environment. Horticultural practices, as well as the growing of plants for landscape, generate great amounts of plastic waste from transplanting pots that are rarely recycled. Nevertheless, there are only few works that deal with biodegradable planting pot preparation and characterization. Planting pots based on biodegradable materials remove the necessity to transplant and hence, discard a container. Planting pots made from industrial and agricultural solid waste, such as wood pulp, paper, or peat moss can be buried directly into the soil altogether with the plant and eventually the container will decompose. Similarly, pots based on biodegradable polymers will also biodegrade when placed in the ground. This chapter reviews the latest findings on biopots (i.e. biodegradable planting pots) based on bioplastics, and also those based on industrial and agricultural waste. Bioplastics usually with addition of different reinforcements, such as plant or wood fibers, are a potential alternative to conventional petroleum-based pots. Also, the use of diverse types of solid residues, such as wood fiber, coconut fiber or coir, rice hull, manure, peat, soil wrap, and straw, in the production chain of novel sustainable products could contribute to the development of modern agriculture. The main thing to consider is that it is necessary to offer rapid biodegradation of planting biopots in soil to avoid their accumulation and root circling while increasing biopots water use efficiency when rising plants. In terms of plant growth and functionality, biodegradable containers can represent a good alternative to replace petroleum-based plastics used for horticulture and floriculture containers.

Keywords

Biopots, Bioplastics, Industrial and Agricultural Waste, Composites

Contents

1. Introduction

In the last six decades, the massive production of plastics has led to an enormous amount of waste worldwide. Over 320 million tons of polymers were produced across the globe in 2015; but unfortunately, less than 10% of the manufactured plastics are actually recycled, and a huge amount is gathering in landfills or thrown away into the environment as litter [1]. Non-renewable petroleum-based plastics are being used in a wide variety of fields. Particularly, modern agriculture uses a huge quantity of plastics stuffs, i.e. direct coverings, greenhouse covering films, soil mulching and solarization films, silage films, shading and protective nets, nets for harvesting and post-harvesting operations, irrigation and drainage pipes, strings and ropes, pots, packaging containers and sacks [2,3].

The practice of using plastic materials in agriculture is often referred to as plasticulture. Even though there are many advantages to the many applications of plastics in agriculture, there are also many concerns. Plastic waste may reduce soil porosity and, therefore, decrease circulation of air. It can also affect microbial populations, and potentially decrease the fertility of the farmland. Fragments of plastic materials have also been shown to release phthalate acid esters into the ground (potentially carcinogenic compounds), where they can be accumulated in crops and, hence, pose a hazard to human health when consumption of contaminated vegetables occurs. Plastics may also accumulate other toxic agrochemicals that are usually applied to crops. This is a special risk for livestock because of their potential to ingest plastic material or other chemicals that leach from it. Plastic pollution can also reach rivers and oceans, which can be toxic for aquatic life [4,5]. Hence, reduction in the use of plastic materials is a major concern in the agriculture field.

One of the most common uses of petroleum-based plastics in agriculture is during transplanting. Transplanting is the practice of removing a plant from its growing place to a different growing site, and it is a process performed all over the world. Farmers and gardeners use pots, containers and cell trays made from a wide variety of materials, with many different shapes, sizes, and colors, to fit different crop species, growing systems, and marketing tactics [6–8]. The vast majority of the pots and trays used for transplanting are based on non-renewable oil-based materials, i.e. polypropylene, polyethylene and polystyrene. These materials have shown desirable mechanical properties, resistance to both chemical and microbial degradation, durability, and also low costs [9]. After their use (usually single-use), oil-based plastic pots lead to contaminated soil, organic matter and agrochemicals. Nonetheless, a correct gathering, discarding and recycling of the used plastic pots have high costs, which leads to their neglecting in landfill or their incineration in a non-controlled manner, with the consequent emission of toxic compounds both into the air and into the ground.

Approximately, 500 million plant containers and seed trays are produced every year. The vast majority are either disposed in landfill or burnt in an uncontrolled manner. A large quantity of fossil fuel is used to manufacture plastic pots, which takes around 500 years to decompose. A feasible substitute to the use of oil-based plastic pots could be the use of biodegradable ones, also known as biocontainers or biopots [7,10–13]. Biopots show technical properties that make them suitable for agricultural applications including their degradation by occurring microorganisms [9,14,15] to inorganic products and biomass, mainly water and carbon dioxide (Fig. 1). Plastics biodegradation is mainly restricted by their chemical structure, molecular weight, solubility in water and their xenobiotic characteristic. Some studies have demonstrated that while some of the bio-based plastics and natural fibers can biodegrade to a substantial degree, plastics with additives that theoretically confer biodegradability to polymers (such as, polypropylene and polyethylene) were not able to significantly enhance the biodegradability of these recalcitrant materials [16].

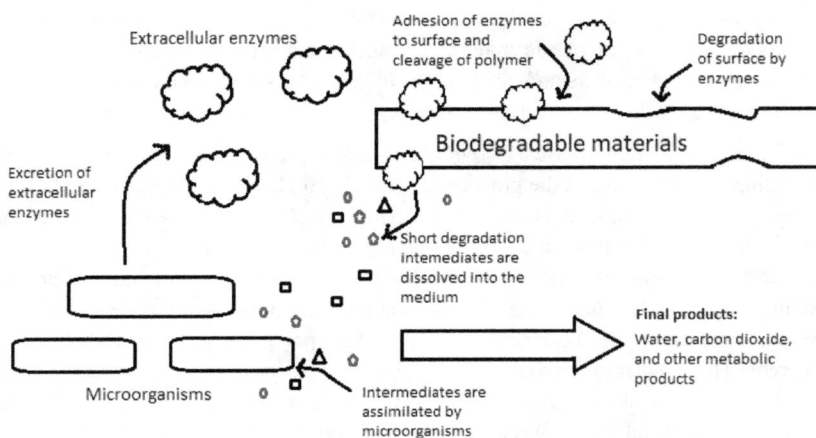

Fig. 1. Mechanism of biodegradable polymers. Scheme adapted from Rydz et al. [17].

Biodegradable containers or biopots are a sustainable alternative to petroleum-based pots that could easily adjust to both horticulture and floriculture production, reducing the enormous amount of plastic waste, and providing outstanding marketing opportunities. Some studies have particularly focus on the marketing aspects of biopots, and have showed that the low appeal of biodegradable pots was mostly due to their appearance. Hall et al. [18] performed a consumer acceptance study which concluded that consumers prefer rice hull containers in the first place, and those made of straw in the second place. Nevertheless, preceding studies performed testing biodegradable containers suggested that these have a significant impact on plant growth [19][20].

This chapter will summarize the latest findings on biopots (i.e. biodegradable planting pots) based on natural and/or synthetic biodegradable polymers, and also those based on industrial and agricultural waste. Their performance on plant growth will also be reviewed.

Advanced Applications of Bio-degradable Green Composites   Materials Research Forum LLC
Materials Research Foundations **68** (2020) 85-103   https://doi.org/10.21741/9781644900659-4

2.   Biodegradable pots (biopots)

Biopots can be classified into two main categories according to their form of use [21]: plantable and compostable (Fig. 2).

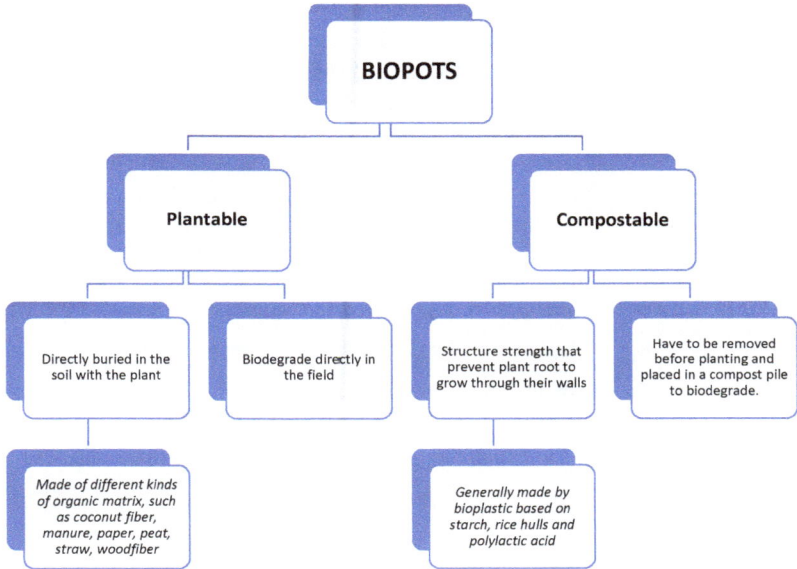

Fig. 2. Classification and main characteristics of biodegradable pots.

Plantable biopots have the main advantage that they can be buried altogether with young plants or seedlings directly into the ground, hence, making the transplanting process much faster, and easier field clean up, as there is no pot disposal (cero waste). Therefore, the use of plantable biopots can reduce farm work effort, cost, and environmental contamination. Plantable biopots can let the roots develop more naturally in the growing media either outdoors into the ground in open field, or indoors in larger containers (e.g. in greenhouse farming, where roots spiraling and binding issues can be avoided). Plantable biopots go through the biodegradation process: once planted in the ground, they transform to produce biomass and inorganic products (i.e. water and carbon dioxide, as previously explained).

These pots lifetime can range from a few months to five years according to their use (outdoors or indoors) [22,23].

On the contrary, compostable biopots are those containers that cannot be buried directly into the soil, but once the transplant has been performed, they can be thrown into a compost or landfill facility where they will biodegrade. Some examples of this kind of biopots include rice hull pots manufactured by Summit Plastic Co. (Ohio, USA), that are based on ground rice hulls with a binder to produce a solid pot; OP47 pots, also produced by Summit Plastic Co., which are biodegradable plastics based entirely on renewable starch produce; and paper-based pots manufactured by Western Pulp Products (Oregon, USA) which are made of recycled paper (more than 74%, with at least 37% post-consumer recycled) and pressed wood pulp. The strong structure of these materials keeps plant roots to develop throughout the pot walls, hence, they cannot be buried altogether with the plant, but need to be first removed and then thrown into a compost pile to let them decompose.

2.1   Biodegradable planting pots based on bioplastics

A biopolymer or bioplastic can be defined by following one of the next criteria:

(1) the source of the raw materials: "bio-based", i.e. made from natural renewable raw materials;

(2) the biodegradability of the polymer.

So, in order to be classified as a bioplastic, the polymer has to either be bio-based or be biodegradable, or both.

In this chapter we will focus on the biodegradable biopolymers, which are defined by the withdrawn standard ASTM D5488-94de1 as polymers that are "capable of undergoing decomposition into carbon dioxide, methane, water, inorganic compounds, or biomass in which the predominant mechanism is the enzymatic action of microorganisms that can be measured by standard tests, over a specific period of time, reflecting available disposal conditions" [24].

Biodegradable polymers can either be fossil fuel-based or bio-based, as it can be seen in Fig. 3. Bioplastics that are bio-based can be either obtained from microorganisms, animals, or plants. It is noteworthy to highlight that bio-based polymers are not sustainable polymers *per se*; the materials sustainability will depend on a variety of issues, such as the source of the material, its manufacture procedure, and the way the product is handled once it has been used.

Advanced Applications of Bio-degradable Green Composites   Materials Research Forum LLC
Materials Research Foundations **68** (2020) 85-103   https://doi.org/10.21741/9781644900659-4

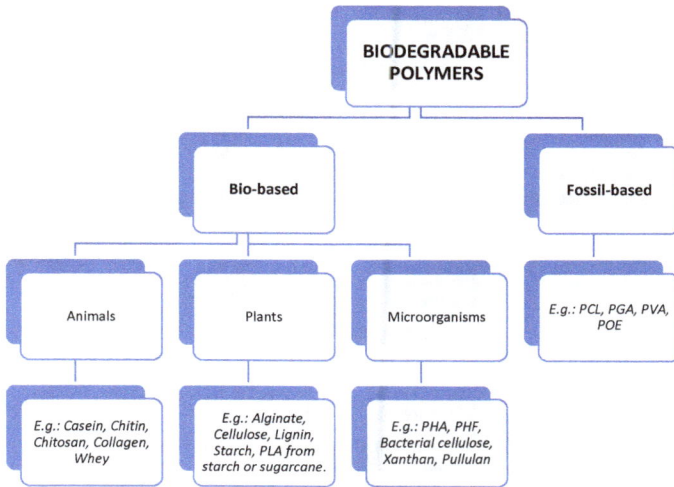

Fig. 3. Classification of biodegradable polymers. PLA: poly(lactic acid); PHA: polyhydroxyalkanoates; PHF: polyhydroxy fatty acid PCL: poly(ε-caprolactone); PGA: poly(glycolic acid; PVA: poly(vinyl alcohol); POE: poly(ortho ester).

Biopolymers or bioplastics, e.g. polylactic acid (PLA) and polyhydroxyalkanoates (PHA), usually show poor mechanical performance, therefore, natural fibers could be used to act as reinforcement enhancing the properties of the matrix (mainly the strength and stiffness), resulting in composite materials with improved performance. Another approach to improve the mechanical properties of biodegradable polymers could be the incorporation of nanoclays [25]. Nanoclay sheets can exfoliate in the biopolymer matrix, resulting in nano-scale particles with an extremely high superficial area and high aspect ratio [26,27]. Examples of different composite biomaterials that have been studied for different agriculture applications are portrayed in Table 1.

Grazuleviciene et al. [34] showed that biodegradable polymeric composites based on starch, peat and polyvinyl alcohol (PVA) are suitable for the production of planting containers. They compared the performance of the developed materials with peat pots showing improvements on mechanical properties, increases on moisture absorption and reduction of vapor permeability, required to maintain steady moisture temperature regime in the soil.

Table 1. Examples of biocomposites for agriculture applications

Matrix	Reinforcement	Application	Reference
Biodegradable polyester	Plant fibers	Pots	[28]
Gutta Percha	Biochar	Films	[29]
Hydrolyzed soybean protein isolate/urea/formaldehyde	Straw fiber	Flowerpots	[30]
PLA	Osage Orange wood particles	Mulch film	[31]
PVA-Corn Starch	Halloysite nanotubes	Mulch film	[32]
Sodium and calcium alginate	Micrometric fibers derived from tomato peel and seeds and from hemp strands	Pots	[33]

PLA: poly(lactic acid); PVA: poly(vinyl alcohol).

Castronuovo et al. [28] have studied three biodegradable containers (biodegradable polyester alone and with plant fibers) and also a conventional pot based on polypropylene (PP). These authors studied those pots performance in two growing cycles of poinsettia (*E. pulcherrima* cv Premium red). Experiments were performed in a heated greenhouse located in southern Italy. The authors analyzed mechanical properties and established that pots made from entirely biodegradable polyester show potential application for the cultivation period of *Poinsettia* for approximately 18 weeks. Nevertheless, the addition of plant fiber to the biodegradable polyester (especially in 20 wt.% of fibers), resulted in a faster degradation, with poor mechanical properties over time, and therefore were not recommended for this use mainly related with some rupture issues during the handling in the marketing stages.

Santos et al. [35] developed biodegradable pots based on Bioplast GS 2189, which is a blend of starch with PLA that shows a stiffness similar to those presented by polystyrene (PS) with thin walls produced by injection processing. These authors concluded that injection processing has a significant effect on the pot properties, inducing that its degradation is mainly related to the shear forces and thermal variations that the polymer is subjected to through the injection procedure. The main objective of this study was to design, develop and characterize a thin walled container for plant germination based on biodegradable material. The performance of biodegradable materials based on PLA and natural fibers was analogous to several traditional plastics used on greenhouse crop production [7].

Promising findings were also stated by Schettini et al. [33], who used sodium and calcium alginate as a binder to obtain pots by using different amounts of natural micrometric fibers derived from tomato peel and seeds, and from hemp strands. Crosslinking with calcium chloride in order to obtain calcium alginate pots led to a more rigid material but with a comparative lower barrier to water vapor and higher water absorption. Additionally, the use of hemp fibers produced an increase in modulus and tensile strength flexural properties [33].

There are a considerably large number of patents regarding the development of planting pots based on bioplastics. For example, US 2009/0025290 A1 entitled "Bottomless plant container" [36] discloses a biodegradable planting pot with a flexible side wall that is stiff enough to support itself. The side wall is based on a biodegradable material made from either PHA or PLA (10–100 wt%) with an optional vegetable load (0–90 wt%) (i.e. starch, flour, cellulose, or a blend of them). This pot can be used for transporting, holding, growing, and planting, either in a flower bed or a field. CA 2688516 A1 entitled "Biodegradable composition, preparation method and their application in the manufacture of functional containers for agricultural and/or forestry use" [37] also divulges the development of a biodegradable pot made of PLA (0-80 wt%) and lignocellulose fibers (i.e. wood and/or grape marc fibers), with lubricating additives, plasticizers, functional additives, compatibilizers, and crystallinity modifiers (0-10%). FR 3014885 A1 also discloses a biodegradable material based on bioplastics [38]. This invention relates to a composition of PLA and/or PHA, further comprising polybutylene succinate (PBS), and at least one type of lamellar nanofillers organically modified. This composition can be used for manufacturing, preferably, a biodegradable flexible packaging.

Hence, bioplastics usually with addition of different reinforcements such as plant or wood fibers, are a potential alternative to conventional petroleum-based pots. In the following section biodegradable pots entirely made from agricultural or industrial waste will be discussed.

2.2   Biopots based on industrial and agricultural waste

Global population, including industry, produces around 1.3 billion tons of solid waste every year, which is projected to double in ten years [39]. Hence, the incorporation of solid waste into the production chain could contribute to the circular economy which has a main goal: to make systems and industrial processes sustainable and environmentally friendly. Therefore, the exploitation of industrial or agricultural waste to obtain new materials is not only convenient for the revalorization of a residue but also to improve yield crops through their incorporation in agricultural systems.

Worldwide, several companies are already producing and commercializing biopots based on a wide variety of solid residues, such as cow manure, coconut husk and wood pulp. Table 2 portrays some examples of commercially available biodegradable pots and trays made from different agricultural and industrial wastes.

Table 2. Commercial biodegradable pots and trays based on industrial and agricultural waste

Manufacturer	Product names*	Composition
CowPots, Connecticut USA	Round CowPot Square CowPot Six Cell CowPot	100% composted cow manure
Fertil, Boulogne-Billancourt France	FERTILPOT FERTILPACK FERTILPOT NT	100% wood fibers
Ivy Acres Inc., New York, USA.	StrawPot	Rice straw, coconut husks and a natural latex adhesive
Jiffy Products of America, Ohio, USA	JiffyPot JiffyStrips	Peat
Summit Plastics Co.	Square Pint NetPot	Slotted rice hull fibers
Summit Plastics Co.	Rice Pot	Solid rice hull
Summit Plastics Co.	EcoGrow	Recycled newspaper and a plant-based binding agent
Summit Plastics Co.	Coconut Coir Pot	Coconut Coir
The HC Companies, Ohio USA	Coconut Coir Pot	Coconut coir
Wester Pulp Products Company	Containers	Molded fibers

As indicated in manufacturers on-line catalogs

Sun et al. [40] studied different containers in field trials. They analyzed seven plantable biocontainers made from different industrial and agricultural waste (i.e. coconut fiber or coir, peat, soil wrap, manure, rice hull, wood fiber, and straw). Field experiments were performed in the United States during both 2011 and 2012 (at five different locations) in order to establish the effect of plantable biopots on plant development and quality, and on the decomposition rate of the pot in landscape. The impact of the type of container on plant growth and quality changed with climate and location, plant species, and growing season.

Lopez & Camberato [41] evaluated quality parameters such as, appearance and durability, of different biopots in a long-term greenhouse crop. They tested wood pulp and *Canadian sphagnum* peat moss, rice straw with coco fiber, composted cow manure, coconut husk, wheat starch-derived bio-resin, rice hull and recycled paper pots using common plastic pot as control. The first four favored algae development and breakage of pots, while the others were more resistant.

Recently, Sartore et al. [13] reported the preparation of biopots from leather industry wastes. Authors have used the protein obtained as by-product of the leather industry in order to obtain bio-pots together with epoxidized soybean oil and ethylene diamine as crosslinking agent. Once cured, the pots were tested in water and disintegrate completely after 2 days.

Another significant factor to mention is the water use efficiency in biopot utilization. Wang et al. [42] reported that bio-based containers presented lower water use efficiency than the traditional ones. These authors analyzed the daily water use of plants grown in non-conventional containers made from recycled paper or coir fiber at different locations in the United States. Their investigation highlights the need for pots that can grow plants of great performance with reduced use of water, since the environmental benefit achieved with the biodegradable pots is not meeting the best water use efficiency [42].

Evans et al. [8] have evaluated nine commercially existing biopots and a plastic control at Arkansas and Mississippi (USA), to study the irrigation cycle and the amount of water needed to produce 'Cooler Grape' vinca (*Catharanthus roseus*) crop with or without the use of plastic shuttle trays. Water-permeable biopots (i.e. coconut fiber, slotted rice hull, wood fiber, straw, manure, and peat pots) loss water through the container walls at higher rates than plastic control containers resulting in more frequent irrigation requirements and more water to produce a vinca crop than the control plastic containers. On the other hand, they have shown that solid rice hull and bioplastic pots (rather impermeable to water) loss water at a similar pace than plastic control ones, having similar water necessities and irrigation cycles. Whereas putting permeable biopots in plastic shuttle trays reduces the

Advanced Applications of Bio-degradable Green Composites   Materials Research Forum LLC
Materials Research Foundations **68** (2020) 85-103   https://doi.org/10.21741/9781644900659-4

loss of water through the container walls; water needs for these biodegradable containers are higher than those for plastic control ones.

Nambuthiri & Ingram [43] reported the effect of biopots on height and coverage of different plant species in comparison with traditional plastic and paper pot. They used a commercial biopot prepared from polyhydroxyl alkanoate polymers obtained by fermenting renewable carbon-based feed stocks, mainly corn, and another biopot made from rice hull. Plants were grown in the different pots and then, the biodegradable ones were planted while the plants in non-biodegradable containers had to be transplanted. Authors reported an important reduction in time when using planting pots instead of transplanting their plants. With regards to their stability, paper and biopolymer pots began to break after 10 weeks of production and degrade almost completely after 4 months in the field. Slow degradation of pots could cause root circling, which could also affect water and nutrient movement and uptake. In turn, the high amount of N and C in soil could retard containers biodegradation in the field [43].

Thus, the use of different kinds of residues in the production chain of novel sustainable products could contribute to the development of a modern agriculture. The main thing to consider is that it is necessary to offer rapid biodegradation of planting biopots in soil to avoid their accumulation and root circling while increasing biopot water use efficiency when rising plants.

3. Effect of biopots on plant growth and quality

Besides the sustainability problems and excessive root temperature especially when the plastic is dark, a functional drawback of using petroleum-based plastic horticulture containers is root circling of some medium- and long-cycle crops. In the case of trees, circling roots can also cause girdling of the stem reducing water and nutrient transport affecting its lifespan [44]. On the other hand, according to several studies, biodegradable pots constitute a good alternative for growing plants since they can be fully decomposable in soil, and also in many cases, bio-based pots improve plant growth and development when compared with traditional non-biodegradable plastic pots. In this sense, Li et al. [45] evaluated the growth of azalea 'Chiffon' (*Rhododendron sp.*) plant in two containers, one made of biodegradable recycled paper and the other of non-biodegradable plastic. Azalea plants were equally fertilized and daily irrigated with equal amount of water. Paper bio-containers produced an increase in plant growth index, leaf weight and area, and total biomass compared with plastic pots. Plants grown in bio-pot showed a significant higher root length and surface area when comparing with those grown in plastic containers, probably resulting from greater evaporative cooling through biocontainer walls, and in turn, lower soil temperature and improved drainage.

Flax et al. [46] compared growth parameters of "Arizona Sun" flower (*Gaillardia ×
grandiflora*) produced in PLA-based container and petroleum-derived polymer pot used
as control. Plant size was generally unaffected by the type of container. Shoots and roots
produced in bioplastic containers were not significantly different to those found in plants
produced in control pots. However, depending on the type and materials, biocontainers
can be damaged or weakened due to environmental conditions during the plant life cycle.
For this reason, since the production times for some woody perennial plants are longer
than that used in this experiment (8–9 weeks) these same authors recommended that next
experiments should be conducted using the biocontainers with crops that span multiple
production seasons. In the case of *Cyclamen persicum* plant, its growth was evaluated on
different biopots, such as solid and slotted rice hull pots, dairy manure, coconut fiber, rice
straw, wood fiber, paper and peat pots using polypropylene pot as control [47]. Notably,
with the exception of wood fiber pot, higher dry shoot and root weights were shown in
plants grown in bio-pots in comparison with plastic ones. However, there were no
significant differences in production time between each container.

Schettini et al. [33] studied the performance of alginate based pots on pepper roots and
height. In all cases, pots presented appropriated mechanical properties to allow material
functionality during 35 days until transplanting degrading in soil in 16 days. The roots of
the pepper seedlings showed good branching structure and also develop secondary
branching in comparison with the ones grown in control pots that showed only long roots.
Also, plants heights in bio-pot with tomato and hemp fiber were significantly higher
(0.76 m) than the ones used as control (0.67 m) [33].

Previously, in 2011 and 2012, seven planting biopots, i.e. manure, coir, peat, wood fiber,
straw, soil wrap, and rice hull pots, were assayed on three plant species: cleome, lantana,
and New Guinea Impatiens. Studies were performed in the United States (in five different
locations). Next, the influence of the different biopots on plant development and quality
was evaluated. Through the first year, the container type did not have a significant impact
on plant growth of any species except in one location. However, in the second year, the
type of container showed a different outcome on plant development according to the
location and species. Curiously, the impact of the type of pot on plant growth was less
considerable than the effect of location. In the same way, plant species has less influence
than the pot material on pot decomposition [8]. In the United States, poinsettias
(*Euphorbia pulcherrima*) are the most valuable potted flower. It is a long-term
greenhouse crop (between 3 and 4 months from transplant to finish). For these reasons,
eight different types of pots, i.e. plastic (or control pot), rice hull, JiffyPot (made from
wood pulp and *Canadian sphagnum* peat moss), StrawPot (based on straw), coconut coir
fiber, CowPot (entirely made of composted cow manure), OP47 Bio (made of wheat

starch-based bioresin), and molded fiber pots were assayed. Appearance and durability of each container were evaluated during the 14-week assay and plants were analyzed at anthesis. Plant heights and root weights from molded fiber were higher than those from OP47 Bio containers; however, shoot dry weigh showed no significant difference among the different pots. Although plant performance was not negatively affected by the studied biopots, the authors concluded that the most promising containers for commercial plant production are molded fiber, rice hull, and OP47 Bio [41]. Otherwise, in agreement with the authors we hypothesize that wall thickness and color might impact on light penetration and in certain conditions, this fact could affect root growth and plant development. As previously mentioned, Lopez and Camberato [41], also tested JiffyPot, CowPot, OP47 Bio containers, rice straw with coco fiber, coconut husk, rice hull and recycled paper pots using common plastic pot as control. Plant growth was improved in recycled paper and rice hull pots over wheat starch-derived bio-resin (OP47), which performance was similar to plastic pot. While there are many other studies related to different types of formulations of biomaterials and biopots used for this type of packaging, we understand that the works discussed here are relevant and novel enough to update us on the subject.

In summary, taking into account all these findings, different biopots can represent innovative alternatives to replace the widely used petroleum-based plastics for horticulture and floriculture containers judged by its similar impact on the establishment of roots and plants accompanied also, by their appropriate sustainable traits for soil and environment.

4.   Future trends and challenges

Modern agriculture uses enormous amounts of non-renewable petroleum-based plastics that are hardly ever recycled. There is a necessity to replace these materials with biodegradable ones, environmentally friendly, that can reduce the negative impact of the former ones. Particularly, pots, containers and seed trays constitute an important segment of the plastic materials used in agriculture, and biodegradable alternatives are gaining an important place in the market. Planting biopots show many advantages to the producer, such as low cost, less farm labor, and no waste, as they are planted together with the seedling or young plant, and biodegraded in the soil directly.

At both laboratory and industrial scales, the use of petroleum-based plastics has been widely valued because of its beneficial properties in the agriculture field. Apparently, biodegradable containers or planting pots based on bioplastics and industrial waste are not less effective than petroleum-based ones. In terms of plant growth and functionality, biodegradable containers can represent a good alternative to replace petroleum-based plastics used for horticulture and floriculture containers particularly, because the use of non-renewable plastic pots is unsustainable. Otherwise, more studies should be performed in order to deepen in the influence of biopots on root architecture before and after transplanting.

References

[1] NatGeo, Planet or plastic? A whopping 91% of plastics isn't recycled., (2018). https://news.nationalgeographic.com/2017/07/plastic-produced-recycling-waste-ocean-trash-debris-environment/ (accessed July 15, 2019).

[2] G. Vox, R.V. Loisi, I. Blanco, G.S. Mugnozza, E. Schettini, Mapping of Agriculture Plastic Waste, Agric. Agric. Sci. Procedia. 8 (2016) 583–591. https://doi.org/10.1016/j.aaspro.2016.02.080.

[3] P. Picuno, Innovative Material and Improved Technical Design for a Sustainable Exploitation of Agricultural Plastic Film, Polym. Plast. Technol. Eng. 53 (2014) 1000–1011. https://doi.org/10.1080/03602559.2014.886056.

[4] J. Wang, S. Lv, M. Zhang, G. Chen, T. Zhu, S. Zhang, Y. Teng, P. Christie, Y. Luo, Effects of plastic film residues on occurrence of phthalates and microbial activity in soils, Chemosphere. 151 (2016) 171–177. https://doi.org/10.1016/J.CHEMOSPHERE.2016.02.076.

[5] R. Scalenghe, Resource or waste? A perspective of plastics degradation in soil with a focus on end-of-life options, Heliyon. 4 (2018) e00941. https://doi.org/10.1016/j.heliyon.2018.e00941.

[6] M.R. Evans, D.L. Hensley, Plant growth in plastic, peat, and processed poultry feather fiber growing containers, HortScience. 39 (2004) 1012–1014.

[7] M.R. Evans, M. Taylor, J. Kuehny, Physical properties of biocontainers for greenhouse crops production, Horttechnology. 20 (2010) 549–555.

[8] M.R. Evans, A.K. Koeser, G. Bi, S. Nambuthiri, R. Geneve, S.T. Lovell, J. Ryan Stewart, Impact of biocontainers with and without shuttle trays on water use in the production of a containerized ornamental greenhouse crop, Horttechnology. 25 (2015) 35–41.

[9] I. Kyrikou, D. Briassoulis, Biodegradation of Agricultural Plastic Films: A Critical Review, J. Polym. Environ. 15 (2007) 125–150. https://doi.org/10.1007/s10924-007-0053-8.

[10] V. Candido, D. Castronuovo, V. Miccolis, The use of biodegradable pots for the cultivation of poinsettia, Acta Hortic. 893 (2011) 1147–1154. https://doi.org/10.17660/ActaHortic.2011.893.132.

[11] S. Kasirajan, M. Ngouajio, Polyethylene and biodegradable mulches for agricultural applications: a review, Agron. Sustain. Dev. 32 (2012) 501–529. https://doi.org/10.1007/s13593-011-0068-3.

[12] E. Schettini, L. Sartore, M. Barbaglio, G. Vox, Hydrolyzed protein based materials for biodegradable spray mulching coatings, Acta Hortic. 952 (2012) 359–366. https://doi.org/10.17660/ActaHortic.2012.952.45.

[13] L. Sartore, E. Schettini, F. Bignotti, S. Pandini, G. Vox, Biodegradable plant nursery containers from leather industry wastes, Polym. Compos. 39 (2018) 2743–2750. https://doi.org/10.1002/pc.24265.

[14] V. Candido, V. Miccolis, G. Gatta, S. Margiotta, P. Picuno, C. Manera, The effect of soil solarization and protection techniques on yield traits of melon in unheated greenhouse, in: Acta Hortic., International Society for Horticultural Science (ISHS), Leuven, Belgium, 2001: pp. 705–712. https://doi.org/10.17660/ActaHortic.2001.559.104.

[15] N. Lucas, C. Bienaime, C. Belloy, M. Queneudec, F. Silvestre, J.-E. Nava-Saucedo, Polymer biodegradation: mechanisms and estimation techniques., Chemosphere. 73 (2008) 429–442. https://doi.org/10.1016/j.chemosphere.2008.06.064.

[16] E.F. Gómez, F.C. Michel, Biodegradability of conventional and bio-based plastics and natural fiber composites during composting, anaerobic digestion and long-term soil incubation, Polym. Degrad. Stab. 98 (2013) 2583–2591. https://doi.org/10.1016/J.POLYMDEGRADSTAB.2013.09.018.

[17] J. Rydz, W. Sikorska, M. Kyulavska, D. Christova, Polyester-based (bio)degradable polymers as environmentally friendly materials for sustainable development, Int. J. Mol. Sci. 16 (2015) 564–596. https://doi.org/10.3390/ijms16010564.

[18] C.R. Hall, R.G. Lopez, J.H. Dennis, C. Yue, B.L. Campbell, B.K. Behe, The appeal of biodegradable packaging to floral consumers, HortScience. 45 (2010) 583–591.

[19] T.J. Hall, W. Lafayette, J.H. Dennis, R.G. Lopez, A.M. Drive, W. Lafayette, M.I. Marshall, W. Lafayette, Factors Affecting Growers ' Willingness to Adopt Sustainable Floriculture Practices, 44 (2009) 1346–1351.

[20] T.J. Hall, J.H. Dennis, R.G. Lopez, M.I. Marshall, Factors affecting growers' willingness to adopt sustainable floriculture practices, HortScience. 44 (2009) 1346–1351.

[21] A. Koeser, S.T. Lovell, M. Evans, J.R. Stewart, Biocontainer water use in short-term greenhouse crop production, Horttechnology. 23 (2013) 215–219.

[22] M. Yamauchi, S. Masuda, M. Kihara, Recycled pots using sweet potato distillation lees, Resour. Conserv. Recycl. 47 (2006) 183–194. https://doi.org/10.1016/J.RESCONREC.2005.10.008.

[23] M. Malinconico, Soil Degradable Bioplastics for a Sustainable Modern Agriculture, 2017. https://doi.org/https://doi.org/10.1007/978-3-662-54130-2.

[24] M. Niaounakis, Definitions of Terms and Types of Biopolymers, in: M. Niaounakis (Ed.), Biopolym. Appl. Trends, William Andrew Publishing, 2015: pp. 1–90. https://doi.org/10.1016/B978-0-323-35399-1.00001-6.

[25] R. Shamsuddin, Protein-Intercalated Bentonite for Bio-composites, University of Waikato, 2013. https://hdl.handle.net/10289/7719.

[26] S.M. Aurebach, K.A. Carrado, P.K. Dutta, Handbook of Layered Materials, 1st Editio, CRC Press, 2004.

[27] S.K. Sharma, A.K. Nema, S.K. Nayak, Effect of modified clay on mechanical and morphological properties of ethylene octane copolymer – polypropylene nanocomposites, (2015). https://doi.org/10.1177/0021998311413686.

[28] D. Castronuovo, P. Picuno, C. Manera, A. Scopa, A. Sofo, Scientia Horticulturae Biodegradable pots for Poinsettia cultivation : Agronomic and technical traits, 197 (2015) 150–156.

[29] D. She, J. Dong, J. Zhang, L. Liu, Q. Sun, Z. Geng, P. Peng, Development of black and biodegradable biochar/gutta percha composite films with high stretchability and barrier properties, Compos. Sci. Technol. 175 (2019) 1–5. https://doi.org/10.1016/J.COMPSCITECH.2019.03.007.

[30] E. Sun, G. Liao, Q. Zhang, P. Qu, G. Wu, H. Huang, Biodegradable copolymer-based composites made from straw fi ber for biocomposite fl owerpots application, Compos. Part B. 165 (2019) 193–198. https://doi.org/10.1016/j.compositesb.2018.11.121.

[31] V.L. Finkenstadt, B. Tisserat, Poly(lactic acid) and Osage Orange wood fiber composites for agricultural mulch films, Ind. Crops Prod. 31 (2010) 316–320. https://doi.org/10.1016/J.INDCROP.2009.11.012.

[32] X. Zeng, B. Zhong, Z. Jia, Q. Zhang, Y. Chen, D. Jia, Halloysite nanotubes as nanocarriers for plant herbicide and its controlled release in biodegradable polymers composite film, Appl. Clay Sci. 171 (2019) 20–28. https://doi.org/10.1016/J.CLAY.2019.01.021.

[33] E. Schettini, G. Santagata, M. Malinconico, B. Immirzi, G. Scarascia Mugnozza, G. Vox, Recycled wastes of tomato and hemp fibres for biodegradable pots: Physico-chemical characterization and field performance, Resour. Conserv. Recycl.

70 (2013) 9–19. https://doi.org/10.1016/J.RESCONREC.2012.11.002.

[34] V. Grazuleviciene, L. Augulis, J. V Grazulevicius, P. Kapitanovas, J. Vedegyte, Biodegradable starch, PVA, and peat composites for agricultural use, Russ. J. Appl. Chem. 80 (2007) 1928–1930. https://doi.org/10.1134/S1070427207110304.

[35] C. Santos, A. Mateus, A. Mendes, C. Malça, Processing and Characterization of Thin Wall and Biodegradable Injected Pots, Procedia Manuf. 12 (2017) 96–105. https://doi.org/10.1016/j.promfg.2017.08.013.

[36] G. Trabka, Bottomless plant container, Patent No. US 2009/0025290 A1, 2009, retrieved from https://patents.google.com/patent/US20090025290.

[37] G. Cabrera Barja, O. Soto Sanchez, Biodegradable composition, preparation method and their application in the manufacture of functional containers for agricultural and/or forestry use, Patent No. CA 2688516 A1, 2010, retrieved from https://patents.google.com/patent/CA2688516A1/fi.

[38] E. Tighzert, H. Thi, L. Nguyen, F. Berzin, S.E. Risse, M. Vitofrancesco, Composition a basic agri-sources and biodegradable polymers, Patent No. FR 3 014 885 A1, 2013, retrieved from https://patents.google.com/patent/FR3014885A1/en.

[39] FCC Environment, Linear to Circular Economy – closing the loop. https://www.fcc-group.eu/en/fcc-cee-group/news-and-media/stories-of-waste/from-linear-to-circular-economy-closing-the-loop.html, 2015 (accessed 11 October 2019).

[40] Y. Sun, G. Niu, A.K. Koeser, G. Bi, V. Anderson, K. Jacobsen, R. Conneway, S. Verlinden, R. Stewart, S.T. Lovell, Impact of biocontainers on plant performance and container decomposition in the landscape, Horttechnology. 25 (2015) 63–70.

[41] R.G. Lopez, D.M. Camberato, Growth and Development of 'Eckespoint Classic Red' Poinsettia in Biodegradable and Compostable Containers, HorTechnology. 21 (2011) 419–423. https://doi.org/10.21273/HORTTECH.21.4.419.

[42] X. Wang, R.T. Fernandez, B.M. Cregg, R. Auras, A. Fulcher, D.R. Cochran, G. Niu, Y. Sun, G. Bi, S. Nambuthiri, R.L. Geneve, Multistate Evaluation of Plant Growth and Water Use in Plastic and Alternative Nursery Containers, Horttechnology. 25 (2015) 42–49. https://doi.org/10.21273/HORTTECH.25.1.42.

[43] S.S. Nambuthiri, D.L. Ingram, Evaluation of Plantable Containers for Groundcover Plant Production and Their Establishment in a Landscape, Horttechnology. 24 (2014) 48–52. https://doi.org/10.21273/HORTTECH.24.1.48.

[44] J.A. Schrader, Bioplastics for Horticulture: An Introduction, in: J.A. Schrader, H.A. Kratsch, W.R. Graves (Eds.), Bioplastic Contain. Crop. Syst. Green Technol.

Green Ind., 2016.

[45] T. Li, G. Bi, R.L. Harkess, G.C. Denny, E.K. Blythe, X. Zhao, Nitrogen Rate, Irrigation Frequency, and Container Type Affect Plant Growth and Nutrient Uptake of Encore Azalea 'Chiffon,' HortScience. 53 (2018) 560–566. https://doi.org/10.21273/HORTSCI12817-17.

[46] N.J. Flax, C.J. Currey, J.A. Schrader, D. Grewell, W.R. Graves, Herbaceous Perennial Producers Can Grow High-quality Blanket Flower in Bioplastic-based Plant Containers, Horttechnology. 28 (2018) 212–217. https://doi.org/10.21273/HORTTECH03922-17.

[47] S.A. Beeks, M.R. Evans, Growth of Cyclamen in Biocontainers on an Ebb-and-Flood Subirrigation System, Horttechnology. 23 (2013) 173–176. https://doi.org/10.21273/HORTTECH.23.2.173.

Advanced Applications of Bio-degradable Green Composites   Materials Research Forum LLC
Materials Research Foundations **68** (2020) 104-137   https://doi.org/10.21741/9781644900659-5

Chapter 5

Biodegradable Pots for Seedlings

S. Barbosa, L. Castillo*

Planta Piloto de Ingeniería Química (UNS – CONICET), Camino La Carrindanga Km. 7, 8000, Bahía Blanca, Argentina

Departamento de Ingeniería Química, Universidad Nacional del Sur, Av. Alem 1253, 8000, Bahía Blanca, Argentina

lcastillo@plapiqui.edu.ar

Abstract

Biodegradable containers are environmentally friendly options to plastic pots commonly used in nursery and greenhouse activities. The use of *plantable* and *compostable* pots based on renewable and natural materials derived from waste or by-products of industrial processes have a potential market to enhance the sustainable character of current production systems. This chapter presents an analysis of the state-of-the-art on the development of biodegradable pots. *Plantable* and *compostable* containers made of different renewable materials, which are studied in the academic field up to patents and commercial products, are presented. Particularly, advantages and disadvantages respect to price, processability, mechanical properties, handling performance, plant quality, water-use efficiency and biodegradability are deeply analyzed and discussed.

Keywords

Compostable Pots, Plantable Pots, Biocomposites, Container Properties, Biodegradability

Contents

1.   Introduction

Seedlings from horticultural and ornamental plants are generally cropped in individual container named pots or cans. However, they are simply called containers in nurseries. These pots can be used for several functions along crop lifecycle such as propagation, growth, transportation, and marketing [1,2,3]. Usually, plastic pots are the most used because of their good mechanical properties, lightweight, durability and low cost. Due to mechanical strength, plastic containers can be handled by automatic machines which are used to fill and seed the crops as well as to transport them [2,3]. It is important to remark that synthetic plastics present comparable mechanical strength both, in dry and wet condition. Moreover, they are not corroded and have a high resistance against mildew and algae growth. Since the use of plastic pots started to be implemented in nursery industry in the 1950's, manufacturers have offered several products, depending on the requirements of the plant growth, shipping conditions or market needs. In this sense, a wide variety of plastic pots are available as commercial products fabricated by injection molding, blow molding and forming by pressure, vacuum or thermally [4]. In this way, these pots present the advantage to be produced in different shapes and sizes and they can be reused and recycled. All these features make plastic pots familiar and convenient for

Advanced Applications of Bio-degradable Green Composites   Materials Research Forum LLC
Materials Research Foundations **68** (2020) 104-137   https://doi.org/10.21741/9781644900659-5

growers, considering also that their massive production provides them a guarantee in obtaining uniform crops.

Even though plastic containers represent a suitable alternative in nursery industry from the economical and mechanical point of view; their use has certain disadvantages in respect to agronomical and ecological issues. In this sense, smooth walls confine the growth of plant roots, favoring their coiling and leading to an ulterior compression on them. When a seedling grows, roots tend to move outwards until they reach the pot walls. At this moment, roots are forced to grow downward, but then the container bottom means another physical barrier to their growth. When roots reach the bottom of a plastic pot, they continue growing to search water and nutrients. The more convenient path to explore the substrate is encircling the pot, resulting in few large roots which started to entangle. They continue wrapping and circling around pot edges over and over and eventually choking themselves. This generates a stress state during plant growth, causing a structural damage since constricted roots impede water and nutrients intake. In addition, plant stem become compressed, favoring tissue damage and limiting even more oxygenation and nutrients input. Moreover, encircling roots around edge pot are more susceptible to heat excess (absence of media for insulation), drought (edges tend to dry first), and disease (unhealth and weak roots). Due to the impermeable character to oxygen of plastic materials, petroleum derived containers allow only roots receive oxygen from the top. Oxygenation is a crucial issue for seedling growth since it favors plant development but also contributes to maintain an adequate temperature in the growing media. In this regard, plastic pots stifle the soil and kill all the beneficial microbiology present in the substrate, even more under summer heat and direct sunlight exposition. For this reason, it is important to take into account the heat stress on plant roots since it is responsible for the crop losses or reduced crop quality as a consequence of high substrate temperature in plastic containers [5,6]. This adverse effect is transmitted to the whole plant, impacting on the aboveground production. When substrate temperature near the roots is high, the following detrimental consequences are observed in plants: i) leaf wilting, chlorosis, and drop; ii) reduction in flower numbers and quality; iii) occurrence of abnormal branching; iv) interference in normal physiological and biochemical processes (e.g., photosynthesis and respiration, water and nutrient uptake, hormone synthesis, and translocation processes), and v) increase of incidence in disease, injury and death [7-9]. Additionally, water impermeability of plastic containers leads to problems of overwatering pot and plant could drown with the resulting development of mold or fungus. It is well known that in order to guarantee a good seedling growth, the root system must be propagated extensively in an early production [10]. When plants are transplanted into landscape, generally they are exposed to different stresses (extreme temperature, mechanical injury and changes in growing environment, among others). All these factors influence negatively on plant growth and can lead to a condition known as "transplant shock" [11, 12]. For this reason, it is crucial that root growth not be committed during the nursery stage, since transplant survival and ulterior growth may later be adversely affected in landscape [13]. This effect proceeds from transplanting where roots are not able to anchor into the soil [14].

Typical plastic pots are made of polyethylene that is not biodegradable, so they must be removed before the plant is transplanted into the soil. Since plastics pots are extensively used an overabundance of unrecovered waste is produced. This fact represents a negative aspect related to the use of plastic pots in nursery and greenhouse activities, which it cannot be associated to the material, but also with human action. The proper final disposition of used containers is directly related to the compromise and responsibility of the human beings. To reduce plastic wastes, two possible solutions have been proposed. One of them is the reuse of plastic containers discarded in greenhouse and nursery activities, but the cost related to the cleaning and sterilizing of used containers makes this alternative unfeasible. Moreover, an additional impediment to this solution is the fact that a great number of containers are generally taken home and disposed by the consumer, without returning to the greenhouse. The other solution is the recycling of plastic containers which allows reducing waste and decreasing energy consumption. Even though recycling programs have grown considerably over the last decades for certain products, recycling of plastic containers from greenhouse and nursery activities is further complicated. This difficulty is based on the presence of organic residues, grease, vegetation, moisture, and pesticide in plastic containers that must be previously eliminated to be recycled [15]. In addition, containers in greenhouses and nurseries are exposed to extreme heat and light conditions which can induce UV light degradation, reducing material recyclability. Since reusing plastic containers is not always possible and recycling options are limited, the industry needs to move in a different direction to reduce plastic waste derived from greenhouse and nursery activities which provides sustainability. An alternative is the use of biopots or biodegradable containers.

In the last few decades, an environmentally friendly awareness has been increased in consumers in several application fields, even those involved in nursery and greenhouse sector. In this sense, "green consumers" have demonstrated interest and commitment for projects or businesses with sustainable goals [16]. Considering previous studies, container type is considered a top factor which has a positive impact on consumer perception [3,17,18]. It is important to mention that this issue has gained more interest than other practices involved in nursery/greenhouse activities, such as organic fertilizer or efficient greenhouse space usage [18]. In this context, the container type is considered more relevant than other purchasing considerations (price, carbon footprint, among others) [3]. All these findings evidence that consumer interests are turning to sustainable pots more than the plants themselves [18]. Thus, green consumer habits have driven to growers to explore practices to make their activity more environmentally friendly in terms of nature impact and public perception [17,19]. Greenhouse and nursery are the main sector interested in the use of biodegradable alternatives (*plantable* or *compostable* containers) to enhance the sustainable character of current production systems. In this context, these biocontainers avoid problems related to its source and disposition of plastic pots and they can be decomposed naturally, eliminating or reducing landfill waste. Additionally, *plantable* containers allow saving labor since they do not need to be removed before planting, which favors their decomposition by microorganisms [4]. Biodegradable containers are fabricated using materials proceeding from renewable sources and/or from by-products of productive activities such as agricultural, animal husbandry, fishing, tannery, etc. [20]. This chapter presents the state of the art on the development of biodegradable pots, analyzing different renewable materials which are studied in the academic field up to patents and commercial products that are available on the market. Advantages and disadvantages in respect to price, processability, mechanical properties, handling performance, plant quality, water-use efficiency and biodegradability are deeply analyzed and discussed.

2. Types of pots

Considering society demands toward greener products or processes, the development and use of biodegradable containers have received great attention in nursery and greenhouse activities. They emerged as substitutes of petroleum-based products which occupy a great landfill extension and can remain in the environment indefinitely, if they are not disposed correctly. The purpose of alternative pots is their own decomposition rather than their contribution to landfill waste. Biocontainers are distinguished from their plastics counterparts in terms of their capability to degrade when they are planted directly into the soil or composted. Under these conditions, biodegradable pots are degraded by bacterial flora producing biomass, methane, water and carbon dioxide, without waste to be disposed. There is a classification of container types based on their requirements and ability to degrade at the end of crop production life. In this sense, containers can be classified as *recycled/recyclable plastic*, *plantable*, or *compostable* [21], as it is shown in Fig. 1. This classification depends on the raw materials and the manufacturing process. There are materials that provide itself the required structural stability and can extended life span for long-term use. These materials allow obtaining *plantable* containers. However, sometimes binding materials are required to give support to container structure such as resins, glue, wax, latex, or cow manure. Depending if binding materials are natural or synthetic, *plantable* or *compostable* pots are obtained.

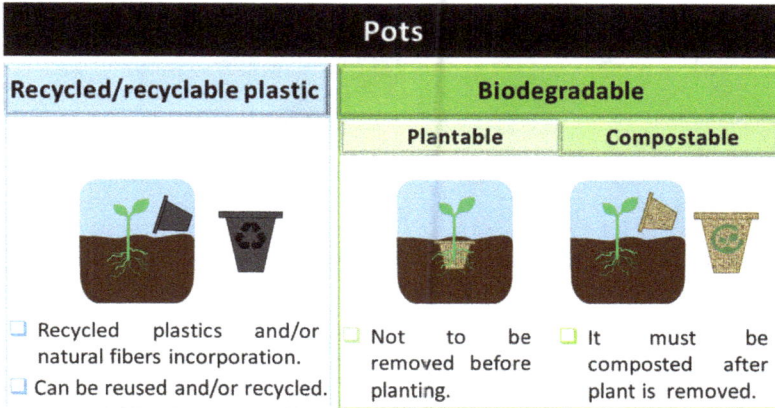

Figure 1. Main types of pots available on the market.

2.1   Recycled and/or recyclable plastic pots

This class of containers includes those made of recycled plastics or their blends with bioplastics and natural fillers. Table 1 shows the advantages and disadvantages associated with recycled/recyclable pots. Generally, they present the advantage to be reused and/or recycled. Although petroleum derived pots are not biodegradable, recycling containers is a step toward sustainability since carbon foot print is reduced compared to conventional plastics pots. These pots are usually made of recycled plastic and natural fibers like cotton, jute, vegetable fibers, or bamboo. They can be obtained with similar shapes and using the same processing operations as conventional plastic pots. For example, plastic and fibers are compounded in molten state. The resulting mixture is thermocompressed, producing a fabric-like geotextile which is sewn into a container [21]. Even though this kind of container is not biodegradable or *compostable,* it is disintegrated slowly due to the natural components, leaving much less residue compared to plastic containers. For this reason, they are referred as alternative containers rather than biodegradable pots.

Table 1. Advantages and disadvantages of recycled/recyclable plastic pots.

Recycled/recyclable plastic pots	
Advantages	**Disadvantages**
❑ Recycled/recyclable	❑ Removal before transplanting
❑ Reusable	❑ Not biodegradable
❑ Mimic the form and function of plastic pots	❑ High substrate temperature

2.2 Plantable pots

This container type enables the direct transplant of seedlings into the ground, reducing the shock associated to this action. *Plantable* pots are distinguished from *compostable* and recycled or bio-based plastic ones since they can be planted into soil, meanwhile the other containers must be externally composted or recycled, respectively. The main requirements considered in the design of *plantable* containers include shipping conditions, watering frequency, and handling. Thus, final pot must last at least one year depending on container material, production practices, and climate conditions, among others. The main function of these biocontainers consists of the quick breaking down once they are planted in order to favor roots penetrating the pot wall and ulterior growing into surrounding soil as containers decompose. This kind of pots is buried into the ground together with the seedling and it is biodegraded as a consequence of the influence of environmental conditions, allowing roots to penetrate the container walls. The biodegradation is also affected by the beneficial interactions between roots and cellulosic microorganisms. This synergistic effect contributes to the recycling and solubilization of mineral nutrients, resulting in an enriched nutrition to plants as a consequence of biodegradable character of the pot. Table 2 presents advantages and disadvantages related to *plantable* pots. *Plantable* containers do not require to be removed before transplanting, reducing the time involved in transplant and landscape cleanup (approx. 20%). Besides the ecological benefits associated to this kind of pots, they also have economic advantages since there are no costs associated with their disposal. Moreover, they reduce the time required for transplanting and cleanup at installation by 20%, which can be a notable advantage to growers. Another benefit of *plantable* pots is their positive contribution to the plant growth since compounds derived from biodegradation are incorporated into the ground and enhance its nutrient content [22]. These containers are a sustainable solution to avoid root disruption and transplanting shock, which are critical problems that occur frequently in plants grown in plastic pots [23]. In this context, it is required that *plantable* containers decompose as fast as possible when they are placed in soil to allow plant establishment, root growth, and future landscape aesthetics and uses [1,24]. However, decomposition rate of these containers must be slow enough to satisfy the requirement of growers during greenhouse production. These containers can be used for ornamental and horticultural plants, allowing a quick and easy transplant. The degradation rate of containers in landscape depends on several factors: i) container material, ii) available nitrogen, iii) moisture, iv) temperature, v) pH, and vi) microbes. Moreover, different results can be obtained with *plantable* containers depending on the geographical region based on soil type and climate. *Plantable* biocontainers allow water passage from roots zone to soil in landscape since most of them are constituted by highly porous materials [25]. In this sense, combinations of peat, manure and wood pulp or paper fibers are employed to obtain these biocontainers. Examples of *plantable* pots are made of coir fiber, wood pulp, rice straw, spruce fibers, rice hulls, wheat, peat, paper, and composted cow manure, as it is shown in Fig. 2. Some biocontainers may draw water from the root zone if they are not properly buried [26]. Moreover, *plantable* pots can absorb additional water respect to plastic ones depending on source materials. In this regard, peat, wood fiber, and manure absorb more water than bioplastics and rice hulls. Another characteristic of *plantable* pots is their contribution to the stabilization of substrate temperature, reducing the probability of root injury mainly in cold regions.

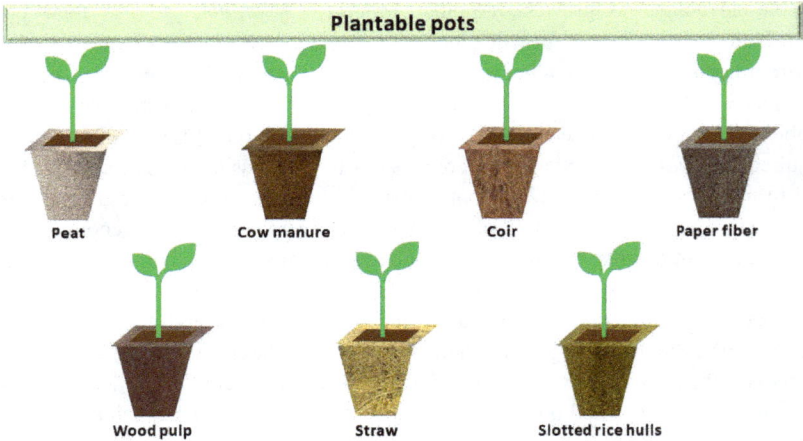

Figure 2. Typical materials used for plantable pots fabrication.

Table 2. Advantages and disadvantages of plantable pots.

Plantable pots	
Advantages	**Disadvantages**
❏ Biodegradable	❏ Increment in water consumption
❏ No removal or disposal costs	
❏ Avoiding root disruption and transplanting shock	❏ Possible breaking during production or transport
❏ Porous structure	❏ Possible delayed break down in soil
❏ Source of nutrients	
❏ Promotion of plant growth	❏ Algae and fungal growth
❏ Stabilization of substrate temperature	

2.3   Compostable pots

When *compostable* containers are used, plants must be removed at the moment of transplant and the containers must be composted separately, either in backyard or using industrial facilities. This kind of container is not degraded quickly or completely when it is placed into soil [27]. Fig. 3 shows the main different materials used for *compostable* pots. Long term containers (lasting approximately 1 year) are made of materials like rice hull, bioplastics, cardboard or wood fibers since they are not quickly degraded. In order to tailor degradation time in commercial *compostable* pots, other materials such as soy, corn, bamboo, poultry feathers, wheat starch and other fiber waste products are incorporated. *Compostable* materials are different from those require industrial composting facilities to break down completely. In this sense, the breakdown of industrially *compostable* containers requires specific conditions of temperature, moisture, pH, aeration and microbial populations, that are not found in a conventional compost pile.

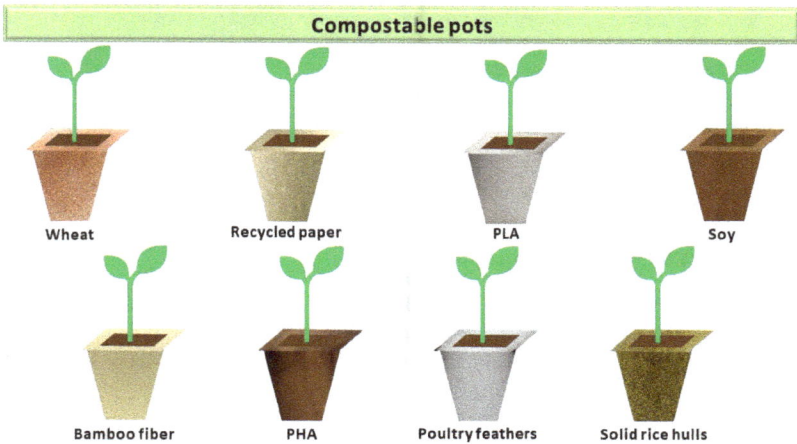

Figure 3. Typical materials used for compostable pots fabrication.

Table 3. Advantages and disadvantages of compostable pots.

Compostable pots	
Advantages	**Disadvantages**
❑ Biodegradable	❑ Removal before transplanting
❑ Porous structure	❑ Be composted separately
❑ Stronger than plantable pots	❑ Not reusable
❑ Influence on plant growth	❑ High drying rate
❑ Lower water loss than plantable pots	❑ Possible breaking during production
❑ Consumer preference for ecofriendly pots	

Each alternative container presents its own characteristics, having advantages or disadvantages depending on the plant requirements, soil type, climate conditions and production practices, among others. In this chapter, special emphasis will be given to biodegradable pots, including *plantable* and *compostable* ones, which are under investigation or are commercially available.

3. Materials for biodegradable pots

A wide variety of materials from renewable and natural sources is used to fabricate biocontainers. Generally, materials derived from waste or by-products of industrial processes can be employed to develop these containers. In this sense, there is a significant reduction in landfill waste since waste is used in containers manufacture.

3.1 Fibers

Several containers obtained from pressed fibers have been investigated and they are available on the market. Different fibrous materials are used for container fabrication such as rice hulls, wheat, paper, peat, wood pulp, spruce fibers, rice hull, coir fiber from coconut palm, rice straw, or bamboo, among others. These containers present a porous structure that favors water and air exchange between roots and surroundings [21]. Source material, moisture content, binder presence and manufacturing process influence on rigidity, strength and decay resistance of pressed fiber containers. In addition, aspects related to production practices such as irrigation frequency, shade/lighting, or temperature could affect pot performance. Moreover, issues associated to plant growth as rooting pattern, pot spacing, and production duration could alter the lifetime of biodegradable containers.

3.1.1 Lignocellullosic residues

Natural fibers and agricultural residues are interesting fillers to be incorporated for biocomposites as it was reported by several authors [20,21,28,29]. Materials derived from plant waste represent good alternatives to help with waste disposal, avoiding the use of synthetic binders [30]. In this sense, lignocellulosic residues have been considered as fibrous raw materials to fabricate biodegradable pots for the seeding in vegetable containerized production. Fibrous materials play a key role in container structure forming, given mainly by their strength properties and biodegradability. In this sense, composites present appropriate mechanical properties to be an adequate pot for containing seedling and substrate along plant development. However, when seedling is transplanted, container has to be disintegrated to avoid affecting the normal growth of the plant. Besides fibrous components, composite materials also included other complements to adjust pot biodegradability, to balance nutritive elements, and to guarantee plant prophylaxis. The most employed lignocellulosic fibers are kraft and recycled pulp to provide the required resistance and consolidate the structure, respectively, and peat since its porous nature favors water inlet and accumulation.

Composites based on peat, cellulosic fibers, residues from grape processing and chemical nutrients, were analyzed as promissory materials for biodegradable containers [31]. Pots from different formulations were obtained through die molding. Fiber presence increased the mechanical strength of pots, meanwhile the incorporation of natural and chemical additives led to the highest rate biodegradation. In this sense, containers reached a biodegradation around 44% at 141 days due to grape residue separates fibers, weakening the pot structure. Regardless composite formulations, planted pots degraded fast compared to those that were not transplanted as a consequence of the rhizosphere effect. Moreover, biodegradation rate during plant growth was different accordingly with the seedling type. For tomato, containers made of peat, cellulosic fibers and chemical aid presented the highest biodegradation rate, meanwhile those containing the residue of grape processing showed the lowest value. When lettuce is planted, the results were reverted. These observations could be related to the fact that tomato roots are denser and more resistant but they grow at a slower rate than lettuce. Consequently, the rhizosphere effect was high in the case of tomato and biodegradation conditions were altered depending on the container type. The presence of complex microorganism populations of heterotrophic bacteria favored the complete decomposition of lignocellulosic material [31].

Cereal bran, a waste derived from flour production, can be an attractive and low-cost filler to be incorporated in polymer composites as raw materials for container development [32]. In this context, bran could be added to extruded papers or pulp containers to control biodegradation rate, as it was reported for different bran-based products [33]. Even though containers made of recycled paper have similar wet and dry strength than those of plastic containers [21], the incorporation of fillers could provide some advantages. In this sense, different formulations based on waste paper, wheat and/or rye bran were prepared to be used in biodegradable pot fabrication [34]. Paper based containers were obtained using a perforated mold where pulp solution and additives (wheat bran or rye bran) were introduced, being water removed by a vacuum pump. Bran incorporation enhanced mechanical performance of paper containers since they are more resistant to damage, compared to commercial pots made of pulp and peat. Moreover, biodegradation rate is influenced by bran content since pots biodegraded more slowly (69 days) than commercial ones (60 days), but faster than containers without fillers (73 days). Container degradation rate is strongly affected by bran amount, exposure time and soil type, although decomposition extent is not influence by filler type (wheat and rye bran). It Is well known that decomposition of paper products is related to microorganism growth. Microbial development could be favored or inhibited, depending on several factors such as soil type, water holding and pH. The lowest degradation rate was observed when paper-based containers were exposed to sandy soil, being degraded after 2 months. A higher extent of biodegradation was observed for pots in forest soil, where complete biodegradation occurred in approximately 1.5 months. Agricultural soil allowed a complete paper decomposition after 1 month, being the most aggressive medium to favor biodegradation pot. Tumer et al. [35] also reported that decomposition process is more severe in organic soil than in sandy one. Since paper-based pots decomposed fast in forest and agricultural soils, they can be suitable as *plantable* containers.

3.1.2 Poultry feathers

Biodegradable containers made of poultry feathers are another alternative which has been studied in the last years. Mixtures of 85% processed feather fiber and 15% kraft paper were prepared to obtain containers [1]. They were used to transplant different plants under several conditions: i) uniform irrigation and fertilization, ii) non uniform irrigation and fertilization; iii) field simulation. Even feather containers were initially hydrophobic, they were be able to absorb water after several irrigation cycles. However, they could not be dried since they would again repel water. These containers allowed that substrates dried more slowly, and water evaporated at a lower rate through the container wall than other biodegradable containers, such as those made of peat. Feather based pots demonstrated to have a differential characteristic given by the additional nitrogen (N) that enables plants growth, both in uniform and non-uniform irrigation and fertilization. The availability of N from feather-based pots favored a higher dry shoot weight of plant than those grown in peat containers. Dry shoot weight of *impatients* and *vinca* grown in feather pots was higher than those grown in plastic or peat containers, under non uniform watering and fertilization. Additionally, results obtained for geranium corroborated the aforementioned behavior under simulated field conditions. Feather based pots did not inhibit root development and plants performed as well those grown in plastic containers, both under simulated field conditions. Concerning water loss, feather pots lost around 2.5 and 3 times lower than peat containers per container and per cm^2, respectively. Plants grown in peat pots needed more water and much more irrigations than those grown in feather containers. Moreover, a higher algal and fungal growth were observed on peat containers than in feather-based pots. The influence of moisture is crucial in container mechanical property. Dry feather pots presented a higher strength in longitudinal direction than dry plastic ones, but lower than dry peat containers. Dry feather and plastic pots showed similar lateral strengths, which are notably higher than dry peat containers. Wet feather containers showed a higher longitudinal strength than peat containers, but a similar strength as plastic pots. However, wet feather containers presented lower lateral strength than wet plastic ones, but higher than wet peat containers. Punch strength for wet feather pots was higher than wet peat pots. Decomposition of feather pot was influenced by the plant species that grown in the container. When tomato was planted, there was no difference on pot decomposition between peat and feather containers. However, feather-based containers were decomposed in a higher extent than peat ones when *vinca* and *marigolg* were planted. In this sense, containers made of feathers present enhanced characteristics respect to those constituted by peat such as the requirement of less water during crop production, higher wet strength, less algae and fungal growth on container walls, and more quick decomposition rate when planted in the field [36].

3.1.3 Residues from solid state fermentation

Biodegradable containers for growing horticultural seedlings have been developed by using a residual substrate derived from the cultivation of medicinal Ganoderma lucidum mushroom [37]. The fabrication process of these containers involved drying, cutting and hollowing of the residual substrate obtained after solid state fermentation of agroindustrial lignocellulosic wastes by the mushroom. In this sense, residual substrates from G. lucidum cultivation on sunflower seed hull, rice straw and rice husk agro-residues were used for making biodegradable containers. The biodegradation during fermentation allowed reducing the cellulose and lignin content, increasing mineralization and supplying nutrients [38,39]. The obtained organic matrix presented good mechanical resistance to bear seedling and plant growth and transplantation. Different assays were performed on these containers to analyze their effect on vegetables germination and seedling quality, as well as, on plant performance from seedling to crop. In this sense, germination of several plant species was not affected, and seedling growth and vigor was enhanced in sunflower seed hull-based containers. Moreover, the porosity of these containers allowed reducing the maximum temperature in the warmest days. In general, the evaluation of containers on seed transplant, plant establishment and fruit yield production revealed that performance of sunflower seed hull-based pots was similar to plastic containers. However, pots made of rice agro-based residue showed an inferior behavior in all the studied stages since did not improve plant growth and vigor and they also affected productivity.

3.1.4 Leather industry wastes

Waste from leather industry is a promissory material to be considered for biodegradable pots since it contains protein hydrolysate (PH) which has an intrinsic agronomic value due to its high nitrogen content [40,41]. Since PH presents a rapid biodegradation, polyepoxy compounds are generally added to increase its environmental durability. Water resistance as well as mechanical and biotechnological behavior of PH was improved by the addition of poly(ethylene glycol) digly-cidyl ether (PEG). PEG is used as a cross-linking and insolubilizing agent since it is water soluble and highly reactive polyepoxy compound [41,42]. However, an environmental alternative is to use epoxidized soybean oil (ESO). In this context, ESO can be used as plasticizer, stabilizer, reactive modifier and diluent [43]. The possibility to use biomaterials based on PH and PEG or ESO in rigid containers and pots as well as in systems for controlled fertilizer release was analyzed [44]. Protein-based aqueous solutions containing dissolved PEG or ESO were prepared. Then, natural fillers such saw dust and wood flour were added to these solutions to adjust material mechanical properties and durability. The resulting suspension was compounded and hot pressed to be ulterior compressed molded to give the final shape to biodegradable pots. Containers obtained from these biocomposites presented enough mechanical strength to assure container functionality during entire lifetime (from seedling to transplant). Moreover, these materials showed a proper resistance to hydrolytic degradation, allowing a slow release of protein [45]. After transplant, biodegradable containers degraded completely favoring root penetration through container walls and plant growth. In respect to commercial plastic pots, biocomposite containers contributed to seedling height and plant growth. The reason of this observation resides on the high nitrogen content present in pH, which is slowly released into the growing media during container degradation, promoting the fertilizing action on soil cultures.

3.1.5 Tomato and hemp fibers wastes

Huge amount of wastes and by-products are generated by agro-food and paper–textile industries, which can be used as source materials for biodegradable containers. In this sense, the negative impact of waste disposal can be reduced since quantity and associated costs are notably diminished. Among these wastes, fibers represent sizeable and functional component. Particularly, natural fibers (jute, kenaf, flax, hemp, among others) as well as those proceed from agriculture residues (stalks of most cereal crops, coconut fibers, peanut shells and tomato seeds and peels, among others) are considered as promissory reinforcing agents for biocomposites [46]. In this context, fibrous components derived from waste of agro-food (tomato peels and seeds) and paper (hemp) industries were proposed to developed composites for biodegradable pots [47]. Containers were obtained from soaking the fibers in a sodium alginate solution that acts as glue. These components were compounded in a blender until obtaining a paste which later was placed into molds to be compressed molded in shaped containers. These biodegradable pots were tested in actual field condition, allowing the root development with a good branching structure. This root system contributed plants to take water and nutrients. At higher tomato fiber content, denser root network and higher seedling height were observed. After being transplanted, these pots completely decomposed in 16 days, allowing the root penetration through containers walls and plant growth [20]. Roots were spread radially and there was no observation of rot or similar symptoms.

3.1.6 Oil extraction residues

Sunflower cake is a residue derived from oil extraction. This material is very rich in proteins that form a dense network, offering thermoplastic properties [48]. Transplanting pots obtained from extruded sunflower cake effectively stimulated the growth of tomato plants [49], even in glycerol presence. On the other hand, as soon as additives such as urea and sulphite of sodium are introduced into the extruded sunflower cake, they cause the death of tomato plants. The first hypothesis to explain this phenomenon is the uptake of oxygen by sodium sulphite, which induces an oxygen deficiency for young plants very sensitive to asphyxiation. The second hypothesis is a toxicity induced by urea: when this nutritive element is too much in the medium, it becomes toxic for plants. Considering these results, horticultural pots were prepared from a base formulation containing 90% extruded sunflower cake in the presence of an aqueous solution of sodium sulfite (TES) and 10% polycaprolactone (PCL) [50]. The incorporation of TES to the formulation makes it easier to bind proteins together. PCL, a synthetic polymer (polyester) of fossil origin, allows increasing the pot durability while maintaining its biodegradability character. The biodegradation rate of these horticultural pots was decreased in the presence of PCL. The purpose to incorporate PCL is to compensate for the addition of extracts that could affect pot mechanical properties of pots, but also offers good water resistance. Injected jars molded from sunflower cake offer time savings on the growth and development of tomato plants before and after transplanting. This is surely linked to an excellent root colonization of the plants in these pots, but perhaps also to a nitrogen supply resulting from the progressive degradation of the protein fraction of the sunflower cake by soil micro-organisms. Although fruit production remains equivalent between plants grown in peat pots and sunflower cake, biomass production remains higher in the case of the latte. Plants are more robust showing wider leaves and larger diameter stems. Their shape and superior mechanical strength for sunflower meal pots compared to peat pots would allow mechanization of transplanting without the need for seed removal, adding value to the product, which can save time for multiple to market gardeners.

3.1.7 Corn husk

Corn husk can be used as material to fabricate biodegradable pots in terms of enhancing sustainability and environmentally friendly character in agricultural activities. Films based on this natural waste was prepared by solution mixing. Biodegradable pots were built from these films which degraded within a period of 7-9 months, reaching the complete decomposition at 270 days [51]. When these pots degrade, water and carbon dioxide are generated without releasing toxic compounds into air or soil. These biodegradable pots can be planted allowing root growth while they decompose.

3.1.8 Banana peels

An alternative waste for making biodegradable containers comes from banana consumption. Since their fibrous character, banana peels seem to be a good candidate to be combined with other natural materials to obtain biopots. Formulations based on tapioca starch were prepared containing different content of banana peels [52]. Biodegradable containers were obtained by molding and their degradability were tested on the ground. Weight loss percentages after 60 days of decomposition in ground revealed that biodegradable pots with lower content of banana peels (50% p/p) were biodegraded in a higher extent (52%) than containers with 70% p/p banana peels (46%). This result was contradictory to the observation by Liew and Khor [53], who reported that bioplastic pot containing higher cellulose content suffered higher weight loss. The behavior of containers made of lower banana peel content could be explained by the lower tensile strength [54]. Even though higher proportion of banana peels increases the microbial decomposition of biocontainers, a lower content of these natural fibers induced a reduction in tensile strength of containers filled with soil, making it highly decomposable [52].

3.2 Bioplastics

Bioplastics are materials derived from renewable sources which have the potential to reach the functional performance of petroleum-based plastics [55]. Blending or compounding these bio-based materials allows achieving a wide range of properties [56]. Bioplastics can be easily degraded by microorganism, producing biomass and carbon dioxide and water [57-59]. In this sense, they have a great potential to be used in the development of containers since they have technical performance suitable for agricultural applications and low environmental footprint. Moreover, they can be employed with potting machines and other automatic equipment, without making substantial changes on machinery. Bioplastics can be used as a coating to enhance the performance of commercial fiber containers. Particularly, coating fiber pots with bioderived polyamide (PA), polylactic acid (PLA), or polyurethane (PUR) allows obtaining an improvement on strength, durability and water-use efficiency during crop production [60]. Besides, health and quality of plants grown in coated containers are improved compared to those grown in uncoated fiber containers. Paper fiber pots coated with PUR are promising products considering their price, processability, performance, plant quality and water-use efficiency. They behave as plastic containers for short- and medium-cycle crops [60]. Another approach is to develop containers from biocomposites where the continuous phase is constituted by biopolymers from renewable and available source [61,62] and the dispersed phase is represented by natural fillers or those coming from waste of agro, textile and food industries [63]. The synergetic combination of these phases results in a material having an enhanced structural, physical and mechanical performance [64,65]. In this sense, natural fibers play a reinforcement role by the improvement of stiffness and strength in composites which can be used for the obtaining of biodegradable containers. Different from conventional fibers (glass, carbon, aramid) which have defined properties, performance of natural fibers is influenced by plant (source and age), separation methods, moisture content, among others [66]. Specific properties (property divided by density) of natural fibers are comparable to those of glass ones since natural fibers have low density [67]. Several works reported the influence of natural fibers on mechanical performance of biodegradable polymers [68,69]. Particularly, chemical similarity between polysaccharides and cellulosic fibers allows obtaining composites with increased tensile strength. The reason of property enhancement is the good fiber/matrix adhesion favored by the highly hydrophilicity of both components [64].

Regarding injection-molded biocomposite containers, fiber addition allows reducing cost and weight, meanwhile improves the processability, function, and biodegradation. Thermoformed containers from wheat-based polymers presented a performance similar to plastic containers respect to water use and growth of geranium [24]. Regarding to assays performed on injection-molded containers, the growth of five common greenhouse species in pots from PLA and soy was healthier, having a better quality than those grown in plastic pots [70]. Additionally, soy-based containers can release nutrients to the soil and improve root morphology by reducing circling and favoring the development of a more fibrous root system. Considering that bioplastics and biocomposites decompose in ground under natural landscape conditions, containers based on these materials are decomposed in organic matter after their life cycle, instead of generating solid waste.

Proteins are being considered as another source of bioplastics. Regardless their origin, animal or vegetable, they are low-cost alternatives for the production of biocontainers. They have high availability on the market since they are byproducts of food and agricultural industry [71]. In this sense, there are some studies focused on zein (protein from corn) as alternative material for biodegradable pots [2,72]. Containers based on zein have agricultural potential since they are easily degradable and compostable without special facilities. Moreover, the hydrophobic character of zein makes it suitable for biodegradable containers since they are exposed to high moisture in greenhouses and nurseries. This property allows extending the stability and durability of zein containers, favoring the feasibility of their commercial use. Prototypes were obtained by molding, using a conventional plastic container. Promissory results were obtained on the release of N from zein based containers which were planted in field soil. Degradation of biocontainers allowed N releasing in landscape, helping the crop growth during production and reducing the application of N fertilizers. Zein containers resulted appropriated for crops which have production cycles lower than 3 months. However, new technologies for zein extraction and recovery should be easier and cheaper in order to make this protein more economically viable than petroleum-based plastics.

Other studies revealed the use of hydrolyzed soy protein isolate (HSPI) as partial replacement of formaldehyde (F) in urea (U)/F adhesives to be used for the obtaining of containers based on straw fiber (SF). Due to good thermal and mechanical properties, biodegradability, and relatively low cost; composite materials made of SF and HSPI/U/F resin are promissory candidates to be used for biodegradable pots fabrication. Hot pressed containers were obtained and then they were buried in a compost soil [73]. Containers of SF reinforced with HSPI/U/F resin were easily decomposed in soil because of microorganism's attack since the macromolecular structure of HSPI/U/F resin was altered.

Gelatin, a protein from animal origin, can be also used to obtain biodegradable pots for seedlings. Gelatin based materials present the advantage to be processed at industrial scale using the same equipment as for thermoplastic polymers. Moreover, low cost, wide availability, as well as, inherent biodegradability makes it a very attractive source material for pots development. In this sense, containers made of gelatin could be directly transplanted into the soil as *plantable* pots. They also could be used to fertilize the surroundings by its own biodegradation and/or by the combination with fertilizers in its formulation. This characteristic confers to gelatin an additional capacity that allows compete economically in the market [41]. Gelatin based formulations containing dissolved urea (as fertilizer) were injected molded to produce biodegradable containers. These materials presented mechanical properties in accordance with container requirements. Since the obtained materials are highly soluble in water, a coating with bee wax was applied to control their biodegradation rate. Once they placed into the ground, gelatin materials enriched with urea could release nitrogen as they were biodegraded. Another alternative to delay gelatin degradation is by the incorporation of natural fillers, which also contributes to increase strength and stiffness of the resulting biocomposite. In this sense, several renewable and natural fillers such as rice hulls, sunflower hulls, and milled leaves from *Ilex paraguariensis* were incorporated into gelatin formulations. Biodegradable containers from these gelatin biocomposites were prepared by molding. These pots containing seedlings were planted into the nursery soil. Biocontainers based on gelatin biocomposite allowed the normal plant grown normally, obtaining higher height than those grown in plastic containers [74].

4.   Commercial biodegradable pots

Global market of pots is mostly dominated by plastic containers. However, biodegradable containers have a great potential. Several efforts to develop biodegradable containers have led to patents that derived on commercial products, which are included in Table 4.

Different studies reported that around 84% of consumers believed that biodegradable containers are more important than plastic ones [3]. Approximately, 73% of consumers considered container type as a driver to purchase rather than price, carbon footprint or waste composition. This implies that certain consumers are willing to pay a high price for an environmentally friendly plant container. Moreover, each kind of biodegradable container has a price premium, which is non uniform for all pot types. In this sense, rice hull-based containers can generate a price premium higher than those of plastic pots, meanwhile the price premium associated to straw and wheat starch containers is not so important [18].

Table 4. Patents related to biodegradable pots.

Patent	Description	Reference
US7681359B2	Biodegradable pot made of straw and coir having at least one surface sprayed with a natural latex.	[75]
US20090249688A1	Biodegradable plant pot based on ply-starch material.	[76]
US8474181B2	Biodegradable plant pot formed from a paperboard-based blank wrapped about an axis.	[77]
US6092331	A planting container constituted by a mat made of natural fibers.	[78]
WO2007020250A1	A plant pot made of paper coated with a polymeric biodegradable.	[79]
DE4009463A1	Planting and/or cultivation pot constituted mainly by pulp corrugated cardboard with a polyester coating.	[80]
CN107736150A	Nutritive flowerpot prepared from pig manure	[81]
JPH10309135A	Nursery biodegradable pot made of plant fibers and a water-soluble binder	[82]
CN1268283A	Seedling pot based on plant fibers	[83]
US20180237655A1	A bio-degradable plant pot having calcium carbonate, cellulosic material, starch and a biodegradable polymer coating.	[84]
CN201393428Y	A degradable flowerpot formed through pressing paper pulp and containing an organic fertilizer	[85]
WO2007051393A1	Naturally degradable green flower pot comprising crushed plant fiber, adhesive and lubricant.	[86]
FR3051097A1	A pot made of bioplastic with a network of biodegradable reinforcement.	[87]

Considering this information, biodegradable containers allow corporations to take advantage of the new product visualization by consumers, but also contribute to the possibility to add high premiums or to increase interest on their products. Another important issue is pot aesthetic which represents an important factor that define the purchase decision. Some biopots are more vulnerable to algae and fungal growth, which could affect negatively on the consumer perception. In this sense, containers made of wood fiber and peat have the highest probability of algae growth, followed by those from manures, paper, and rice straw [24]. Another important uncertainty associated to biodegradable containers is the concern about the possibility to hinder the plant growth during the greenhouse stage. Additionally, there is a dilemma concerning the feasibility of biocontainers to be used in modern mechanized operations. This is one of the reasons for which growers have not yet completely changed to biocontainers implementation. Some biodegradable containers have high dry strength, even higher than those of plastic one; however wet strength in biopots is much lower than those determined at dry condition [24]. In this sense, containers made of wood fiber, peat, and manure present extremely low wet strength that impedes their manual handling.

Biodegradable pots are already produced by worldwide companies. Table 5 reports different commercial biocontainers that are available on market. These products have a life-time which ranges from few months to 1–2 years.

Commercial biodegradable containers have been studied by many authors in order to analyze comparatively their physical properties, greenhouse and landscape performance, biodegradation, plant growth, among others. Evans et al. [24] analyzed different types of commercial containers comprising plastic, wood fiber (Fertil), cow manure (Cowpot), coconut fiber (ITML Horticultural Products), peat (Jiffy), rice hull (Summit Plastic Co.), paper (Western Pulp Products), and rice straw. In this sense, physical properties of biodegradable pots were different from those of plastic ones. Moreover, several types of biocontainers presented dissimilar properties among them. On the other hand, wet strength and water use are relevant topics to be considered when choosing a biocontainer. Rice hull pots presented the highest vertical dry strength of containers tested by Evans et al. [24]. Respect to plastic pots and paper ones, they had similar vertical dry strength. Wood fiber, cow manure and peat containers presented lower vertical dry strength than plastics and paper pots but had higher values than those corresponding to coconut fiber and rice straw containers. Concerning lateral dry strength, rice hull and paper pots showed the highest value. Meanwhile, containers made of rice straw, cow manure, and plastic demonstrated to have comparable strength among them at dry state. Also, these values were superior to those obtained for wood fiber, peat and coir pots. Plastic and cow manure containers presented the highest punch resistance at dry state, meanwhile rice straw, coir and peat pots showed the lowest values for this property.

Table 5. Commercial biodegradable containers.

Product name	Container type	Manufacturer
Biodegradable Pot	Bamboo fiber	Kingfree International Ltd (https://kingfree.en.ec21.com/)
Coir Fibre Pot	Coconut fiber	Coco Pots (https://www.coir-pots.com/)
CowPot	Manure	CowPots (https://cowpots.com/)
Jiffy-Pot®	Peat	Jiffy (http://www.jiffypot.com/en.html)
Fertil	Wood fiber	Fertil (https://www.fertil.fr/pots-biodegradables/)
Fertil pots	Wood fiber	Fertil USA (http://www.fertil.us/)
Biodegradable pots	Bamboo pulp, rice straw, wheat straw	Enviroarc (http://www.enviroarc.net/)
Fertil pot	Wood fiber and peat	Projar (https://www.projar.es/)
Biodegradable Coir Pots	Coir fiber	The Natural Gardener (https://www.thenaturalgardener.co.uk/biodegradable_coir_pots.php)
Biodegradable NFT Pots	Natural fibers (not specified)	Pure Hydrophonics (https://purelyhydroponics.com/about-us/)
Eco 360 Net	Rice hull	Summit Plastic Co (http://www.summitplastic.com/)
Eco 360	Rice hull	Summit Plastic Co (http://www.summitplastic.com/)
Ecoforms pot	Rice hull and starch	Ecoforms (http://ecoforms.com/)
Ellepot	Paper (sleeves)	Ellepot (https://www.ellepot.com/)
6X6RD, 7X7RD	Recycled paper and/or cardboard	Western Pulp (https://www.westernpulp.com/)
KordFibergrow	Recycled paper or cardboard	The HC Companies (https://hc-companies.com/)
Coir pot	Coconut fiber	The HC Companies (https://hc-companies.com/)

Regarding wet vertical and lateral strengths, the highest values were determined for rice hull, paper, and plastic pots. The latter also resulted in higher punch resistant at the wet state, meanwhile the lowest values were measured for containers made of wood fiber, peat, and cow manure. To evaluate the amount of water that is lost per surface area, a substrate previously saturated with water was introduced into each pot, recovering the drainage holes and substrate with wax [24]. Rice hull and plastic containers lost water at the lowest rate while peat, rice straw and wood fiber pots presented the highest values. Plant growth and water use depend strongly on container type. Commercial pots based on solid rice hull demonstrated to have the best performance in respect to total water consumption compared to other biocontainers (coconut fiber, peat, wood fiber, wheat, straw, cow manure, slotted rice hull). Moreover, pots made of a solid rice hull allowed a normal growth of plants. On the other hand, a decreased dry weight is observed for containers based on wood fiber, straw, and manure, which materials have a fast drying rate [25]. Water requirement of commercial containers was evaluated by planting geranium (*Pelargonium ortorum*) in each type of pot. Coir, peat, rice straw, cow manure, and wood fiber containers needed a higher water amount and shorter irrigation intervals than plastic pots. Rice hull containers required similar amount of water and irrigation intervals as plastic pots to produce a geranium crop. Plant growth of *geranium, impatiens,* and *vinca* was different depending on container type and location. Plants grown in plastic or paper pots performed better than those grown in other biodegradable containers. However, plant growth in containers used in the greenhouse produced marketable plants for both the retail and landscape markets [26].

Generally, the less-processed containers such as those made of dairy manure, peat, and wood pulp, favored algae growth [88]. Wood fiber and peat containers were covered by algae and fungal biofilm in 26% and 47%, respectively; meanwhile the other tested pots presented 4% of covering. Concerning biodegradation, cow manure pots presented a different decomposition extent after 8 weeks depending on the location (48% in Louisiana and 62% in Pennsylvania). Biodegradation of wood fiber, rice straw, and peat containers in Pennsylvania reached 24%, 28%, and 32%, respectively; meanwhile in Louisiana percentages were 2%, 9%, and 10%, respectively. The lowest decomposition level was observed for coconut fiber containers (4% in Pennsylvania and 1.5% in Louisiana).

In general, all commercial biocontainers are compatible with automatic machines. Particularly, coir and paper-based containers showed certain difficulties to be handled by machinery, in comparison with plastic ones. However, this drawback can be solved by selecting or developing an appropriate spacing equipment. Additionally, some biodegradable pots such as those made of pressed manure and peat, can suffer damage

Advanced Applications of Bio-degradable Green Composites   Materials Research Forum LLC
Materials Research Foundations **68** (2020) 104-137   https://doi.org/10.21741/9781644900659-5

during shipping which could be a major concern for growers if they cannot be sold. Considering the issue of plant growth, biodegradable containers appear to be proper candidates to replace plastic pots, even though the need for using different irrigation methods [25].

The influence of commercial container type on plant performance depend on climate, growing season, and species. Container type has little influence compared to the impact of climate on plant performance. Even though container degradation was mainly affected by pot composition, climate and species influenced in a lesser extent. After 3 to 4 months in the field, the highest decomposition rate was observed for manure pots (88%) for different locations in United States and two growing seasons. Decomposition percentages of rice hull, coconut fiber, peat, soil wrap, wool fiber and straw containers were 18%, 25%, 38%, 42%, 46%, and 47%, respectively [89]. *Plantable* containers allowed plant establishment as well as plant growth after transplant. Pot material showed a greater influence on container biodegradation than plant species.

Depending on the crop, location as well as cultural conditions and practices; some properties or characteristics of commercial biodegradable containers will be more or less relevant than others. In this sense, considering the specific physical properties of biocontainers, greenhouse managers and landscapers must decide which of them are most important in order to choose the proper container that best match their needs. However, there are a lot of alternatives to plastic containers that can be used successfully for greenhouse production and landscape planting based upon production practices and landscape requirements.

5.   Future trends

Biodegradable pots are an environmentally friendly alternative to plastic pots commonly used in the nursery and greenhouse sector. The green industry promotes the wide spreading of the use of these kinds of containers supported by the fact that each year different options are available on the market. Pot development and involved manufacturing technologies are still evolving and upgrading according to even more increasing grower and consumer acceptance. These biodegradable pots must satisfy the primary functions such holding the growing media, draining, a healthy root development, but also it must guarantee to not alter the root system after planting them into soil. The study and comprehension of the influence of pot properties and nursery operations on plant health and growth allows defining the biodegradable container type according to the requirements. Formulation and dimensions of containers affect root development, availability of both water amount and mineral nutrients, nursery layout, bench size, production scheduling, and plant transportation method. Depending on plant requirements, root system morphology, target plant criteria and economics, there is a more appropriate option for choosing biodegradable pot. Considering a wide variety of plant species are grown in most nurseries, several pots types could be required for each specie in particular.

Even though biodegradable pots tend to be more expensive than plastic ones, consumers are willing to pay more for non-plastic alternatives. This offer a huge demand for biodegradable pots. Additional costs of biodegradable pots could justify their used by growers for value-added products. Moreover, if pot production would be leveled to industrial scale, costs can be lowered. Additionally, there is a huge opportunity to lower the involved costs promoting the fabrication of *plantable/compostable* pots by using waste or recycled products. The increasing demand for biodegradable containers observed in the last years seems not to be decreased, allowing reducing even more the costs associated to pots. Consequently, advantages from agronomical, ecological, mechanical and economical point of view could induce changes in conventional production techniques in the future.

Proper selection of biodegradable pot type for specific application could be assessed through the information provided by academic research and commercial products available on the market, including containers for horticulture, ornamental or forest nurseries. In this sense, the design of biodegradable pot can be tailored according the degradation rate required to crop cycle duration or specific types of soils, among others.

Even though the study and development of *plantable* and *compostable* pots are gaining importance, there is still much to know about them. Despite the influence of biodegradable containers on plant growth, water consumption, economic and environmental issues have been partially studied, several issues need to be considered. Biodegradable containers offer a great opportunity to provide additional benefits by incorporating different compounds in their formulation since they can be released while container is biodegraded. In this way, recent developments comprise biodegradable containers impregnated with natural color, or including slow releasing fertilizers, fungicides, insecticides and plant growth regulators. These innovative ecofriendly products are gaining entry in the market and they could improve the efficiency of the production system. Researchers and industrials must continue to work together, focusing on the development of tailored ecofriendly biodegradable containers. The challenge is to reach emerging characteristics and properties required by growers and customers that will be compatible with current production practices and economically feasible.

References

[1] M.R. Evans, D.L. Hensley, Plant growth in plastic, peat, and processed poultry feather fiber growing containers, Hort. Sci. 39(5) (2004) 1012-1014. https://doi.org/10.21273/HORTSCI.39.5.1012

[2] M.S. Helgeson, W.R. Graves, D.S. Grewell, G. Srinivasan, Degradation and nitrogen release of zein-based bioplastic containers, J. Environ. Hort. 27 (2009) 123-127.

[3] C.R. Hall, B.J. Campbell, B.K. Behe, C. Yue, R.G. Lopez, J.H. Dennis, The appeal of biodegradable packaging to floral consumers, Hort. Sci. 45(2010) 583-591. https://doi.org/10.21273/HORTSCI.45.4.583'po

[4] M. Chappell, G.W. Knox, Alternatives to petroleum-based containers for the nursery industry, Bulletin 1407, University of Georgia, Cooperative Extension 2012.

[5] H. Mathers, Pot-in-pot container culture, The Nurs. Pap. 2 (2000) 1-6.

[6] J.W. Markham, D.J. Bremer, C.R. Boyer, K.R. Schroeder, Effect of container color on substrate temperatures and growth of red maple and redbud, Hort. Sci. 46 (2011) 721–726. https://doi.org/10.21273/HORTSCI.46.5.721

[7] D.L. Ingram, C. Martin, J. Ruter, Effect of heat stress on container-grown plants, Comb. Proc. Int. Plant Propagators Soc. 39 (1989) 348-353.

[8] T.G. Ranney, M.M. Peet, Heat tolerance of five taxa of birch (Betula): Physiological responses to supraoptimal leaf temperatures, J. Am. Soc. Hortic. Sci. 119(2) (1994) 243-248. https://doi.org/10.21273/JASHS.119.2.243

[9] J.E. Webber, S.D. Ross, Flower induction and pollen viability for western larch, Report from U.S. Department of Agriculture, Forest Service, Intermountain Research Station, 1995.

[10] H. Davidson, R. Mecklenburg, C. Peterson, Nursery management: Administration and culture, 4th Edition, Prentice Hall Upper Saddle River, 2000.

[11] H.M. McKay, A review of the effect of stresses between lifting and planting on nursery stock quality and performance, New Forest. 13(1-3) (1996) 363-393.

[12] A.K. Koeser, J.R. Stewart, G.A. Bollero, D.G. Bullock, D.K. Struve, Impacts of handling and transport on the growth and survival of balled-and-burlapped trees, Hort. Sci. 44(1) (2009) 53-58. https://doi.org/10.21273/HORTSCI.44.1.53

[13] L.E. Richardson-Calfee, J.R. Harris, R.H. Jones, J.K. Fanelli, Patterns of root production and mortality during transplant establishment of landscape-sized sugar

maple, J. Am. Soc. Hortic. Sci. 135(3) (2010) 203-211.
https://doi.org/10.21273/JASHS.135.3.203

[14] J. Muriuki, A. Kuria, C. Muthuri, A. Mukuralinda, A. Simons, R. Jamnadass,
Testing biodegradable seedling containers as an alternative for polythene tubes in
tropical small-scale tree nurseries, Small Scale For. 13(2) (2014) 127-142.
https://doi.org/10.1007/s11842-013-9245-3

[15] J.W. Garthe, P.D. Kowal, Recycling used agricultural plastics, Penn State Fact
Sheet C-8, http://pubs.cas. psu.edu/freepubs/pdfs/C8.pdf, 1993.

[16] C. Yue, C. Tong, Organic or local: Investigating consumer preference for fresh
produce using a choice experiment with real economic incentives, Hort. Sci. 44(2)
(2009) 366-371. https://doi.org/10.21273/HORTSCI.44.2.366

[17] J.L. Dennis, R.G. Lopez, B.K. Behe, C.R. Hall, C. Yue, B.L. Campbell, Sustainable
production practices adopted by greenhouse and nursery plant growers, Hort. Sci. 45
(2010) 1232-1237. https://doi.org/10.21273/HORTSCI.45.8.1232

[18] C. Yue, J.H. Dennis, B.K. Behe, C.R. Hall, B.L. Campbell, R.G. Lopez,
Investigating consumer preferences for organic, local, or sustainable plants, Hort. Sci.
46 (2011) 610-615. https://doi.org/10.21273/HORTSCI.46.4.610

[19] T.J. Hall, J.H. Dennis, R.G. Lopez, M.I. Marshall, Factors affecting growers'
willingness to adopt sustainable floriculture practices, Hort. Sci. 44 (2009) 1346-1351.
https://doi.org/10.21273/HORTSCI.44.5.1346

[20] E. Schettini, G. Santagata, M. Malinconico, B. Immirzi, G.S. Mugnozza, G. Vox,
Recycled wastes of tomato and hemp fibres for biodegradable pots: Physico-chemical
characterization and field performance, Resour. Conserv. Recy. 70 (2013) 9-19.
https://doi.org/10.1016/j.resconrec.2012.11.002

[21] S. Nambuthiri, A. Fulcher, A.K. Koeser, R. Geneve, G. Niu, Moving toward
sustainability with alternative containers for greenhouse and nursery crop production:
A review and research update, Hort. Technol. 25(1) (2015) 8-16.
https://doi.org/10.21273/HORTTECH.25.1.8

[22] Abaecherli, V.I. Popa, Lignin in crop cultivations and bioremediation, Environ. Eng.
Manag. J. 4(3) (2005) 273-292. https://doi.org/10.30638/eemj.2005.030

[23] A.A. Khan, T. McNeilly, C. Collins, Accumulation of amino acids, proline, and
manganese stress in maize, J. Plant. Nutrition. 23 (2000) 1303-1314.
https://doi.org/10.1080/01904160009382101

[24] M.R. Evans, M. Taylor, J. Kuehny, Physical properties of biocontainers for greenhouse crops production, Hort Technol. 20 (2010) 549-555. https://doi.org/10.21273/HORTTECH.20.3.549

[25] Koeser, R. Hauer, K. Norris, R. Krouse, Factors influencing long-term street tree survival in Milwaukee, WI, USA, Urban For. Urban Gree. 12(4) (2013) 562-568. https://doi.org/10.1016/j.ufug.2013.05.006

[26] J.S. Kuehny, M. Taylor, M.R. Evans, Greenhouse and landscape performance of bedding plants in biocontainers, Hort. Technol. 21 (2011) 155-161. https://doi.org/10.21273/HORTTECH.21.2.155

[27] B.P. Mooney, The second green revolution? Production of plant-based biodegradable plastics, Biochem. J. 418 (2009) 219-232. https://doi.org/10.1042/BJ20081769

[28] S. Ochi, Durability of starch based biodegradable plastics reinforced with manila hemp fibers, Materials 4 (2011) 457–468. https://doi.org/10.3390/ma4030457

[29] T. Tesfaye, B. Sithole, D. Ramjugernath, V. Chunilall, Valorisation of chicken feathers: Application in paper production, J. Clean Prod. 164 (2017) 1324–1331. https://doi.org/10.1016/j.jclepro.2017.07.034

[30] C. Müller, U. Kües, C. Schöpper A. Kharazipour, Natural binders, in: U. Kües (Ed.), Wood production, wood technology, and biotechnological impacts, Universitätsverlag Göttingen, Göttingen, 2007, pp. 347-381.

[31] P. Nechita, E. Dobrin, F. Ciolacu, E. Bobu, The biodegradability and mechanical strength of nutritive pots for vegetable planting based on lignocellulose composite materials, Bio Resources 5(2) (2010) 1102-1113.

[32] K. Formela, A. Hejna, Ł. Piszczyk, M.R. Saeb X. Colom, Processing and structure–property relationships of natural rubber/wheat bran biocomposites, Cellulose 23 (2016) 3157–3175. https://doi.org/10.1007/s10570-016-1020-0

[33] Sandak, I. Modzelewska, J. Sandak, FT-NIR analysis of recycled paper with addition of cereal bran biodegraded with microfungi, J. Near Infrared Spectrosc. 19(5) (2011) 369–379. https://doi.org/10.1255/jnirs.951

[34] Sandak, J. Sandak, I. Modzelewska, Manufacturing fit-for-purpose paper packaging containers with controlled biodegradation rate by optimizing addition of natural fillers, Cellulose 26 (2019) 2673–2688. https://doi.org/10.1007/s10570-018-02235-6

[35] A.R. Tumer, E. Karacaoglu, A. Namli, A. Keten, S. Farasat, R. Akcan, O. Sert, A.B. Odabasi, Effects of different types of soil on decomposition: an experimental study, Leg. Med. 15(3) (2013) 149–156. https://doi.org/10.1016/j.legalmed.2012.11.003

[36] M.R. Evans, D. Karcher, Properties of plastic, peat, and processed poultry feather fiber growing containers, Hort. Sci. 39 (2004) 1008-1011. https://doi.org/10.21273/HORTSCI.39.5.1008

[37] P.D. Postemsky, P.A. Marinangeli, N.R. Curvetto, Recycling of residual substrate from Ganoderma lucidum mushroom cultivation as biodegradable containers for horticultural seedlings, Sci. Hortic. Amsterdam. 201 (2016) 329–337. https://doi.org/10.1016/j.scienta.2016.02.021

[38] P.D. Postemsky, S.E. Delmastro, N.R. Curvetto, Effect of edible oils and Cu (II) on the biodegradation of rice by-products by *Ganoderma lucidum* mushroom, Int. Biodeter. Biodegr. 93 (2014) 25-32. https://doi.org/10.1016/j.ibiod.2014.05.006

[39] P.D. Postemsky, N.R. Curvetto, Solid-state fermentation of cereal grains and sunflower seed hulls by *Grifolagargal* and *Grifolasordulenta*, Int. Biodeter. Biodegr. 100 (2015) 52-61. https://doi.org/10.1016/j.ibiod.2015.02.016

[40] E. Chiellini, P. Cinelli, R.S. Kenawy, A. Lazzeri, Gelatin-based blends and composites. Morphological and thermal mechanical characterization, Biomacromolecules 2 (2001) 806-811. https://doi.org/10.1021/bm015519h

[41] L. Sartore, G. Vox, E. Schettini, Preparation and performance of novel biodegradable polymeric materials based on hydrolyzed proteins for agricultural application, J. Polym. Environ. 21(3) (2013) 718–725. https://doi.org/10.1007/s10924-013-0574-2

[42] H. Kono, Characterization and properties of carboxymethyl cellulose hydrogels crosslinked by polyethylene glycol, Carbohydr. Polym. 106 (2014) 84-93. https://doi.org/10.1016/j.carbpol.2014.02.020

[43] P. Niedermann, G. Szebényi, A. Toldy, Effect of epoxidized soybean oil on curing, rheological, mechanical and thermal properties of aromatic and aliphatic epoxy resins, J. Polym. Environ. 22(4) (2014) 525-536. https://doi.org/10.1007/s10924-014-0673-8

[44] L. Sartore, E. Schettini, S. Pandini, F. Bignotti, G. Vox, A. D'Amore, Biodegradable containers from green waste materials, in AIP Conference Proceedings. (Vol. 1736, No. 1, p. 020100), AIP Publishing, 2016. https://doi.org/10.1063/1.4949675

[45] L. Sartore, F. Bignotti, S. Pandini, A. D'Amore, L. Di Landro, Green composites and blends from leather industry waste, Polym. Composite. 37(12) (2016b) 3416-3422. https://doi.org/10.1002/pc.23541

[46] P.A. Sreekumar, A. Pradeesh, G. Unnikrishnan, J. Kuruvilla, T. Sabu, Mechanical and water sorption studies of ecofriendly banana fiber reinforced polyester composites fabricated by RTM, J. Appl. Polym. Sci. 109 (2008) 1547–1555. https://doi.org/10.1002/app.28155

[47] Ashori, A. Nourbakhsh, Reinforced polypropylene composites: Effects of chemical compositions and particle size, Bioresour. Technol. 101(7) (2010) 2515-2519. https://doi.org/10.1016/j.biortech.2009.11.022

[48] M.C. Geneau, Proceded'elaborationd'agromateriau composite naturel par extrusion bivis et injection moulage de tourteau de tournesol, PhD thesis, Institut National Polytechnique De Toulouse, 2006.

[49] Rouilly, F. Silvestre, L. Rigal, H. Caruel, E. Paux, J. Silvestre, P. Morard, Utilisation de tourteau de tournesol pour la fabrication de pots de repiquage biodégradables, 15th International Sunflower Conference, Toulouse, France, 2000.

[50] M. C. Celhay, Fractionnement de coproduits de pin maritime (Pinus pinaster) et de peuplier (Populustremula) pour l'obtentiond' extraits polyphénoliques à activité antioxydante: procédéd'extraction aqueuseen extracteur bi-vis et étude des conditions subcritiques, PhD thesis, Université de Toulouse, 2013.

[51] M.Z. Norashikin, M. Z. Ibrahim, The potential of natural waste (corn husk) for production of environmental friendly biodegradable film for seedling, World Acad. Sci. Eng. Technol. 58 (2009) 176-180.

[52] S.N.A.M. Rafee, Y.L. Lee, M.R. Jamalludin, N.A. Razak, N.L. Makhtar, R.I. Ismail, Effect of different ratios of biomaterials to banana peels on the weight loss of biodegradable pots, Acta Technologica Agriculturae. 22(1) (2019) 1-4. https://doi.org/10.2478/ata-2019-0001

[53] K.C. Liew, L.K. Khor, Effect of different ratios of bioplastic to newspaper pulp fibres on the weight loss of bioplastic pot, J. King Saud Univ. Eng. Sci. 27 (2015) 137–141. https://doi.org/10.1016/j.jksues.2013.08.001

[54] Ververis, K. Georghiou, N. Christodoulakis, P. Santas, R. Santas, Fiber dimensions, lignin and cellulose content of various plant materials and their suitability for paper production, Ind. Crop. Prod. 19 (2003) 245–254. https://doi.org/10.1016/j.indcrop.2003.10.006

[55] Grewell, G. Srinivasan, J. Schrader, W. Graves, M. Kessler, Sustainable materials for a horticultural application, Plast. Eng. 70(3) (2014) 44-52. https://doi.org/10.1002/j.1941-9635.2014.tb01141.x

[56] S.A. Madbouly, J.A. Schrader, G. Srinivasan, K. Liu, K.G. McCabe, D. Grewell, W.R. Graves, M.R. Kessler, Biodegradation behavior of bacterial-based polyhydroxyalkanoate (PHA) and DDGS composites, Green Chem. 16(4) (2014) 1911-1920. https://doi.org/10.1039/C3GC41503A

[57] V. Candido, V. Miccolis, G., Gatta, S., Margiotta, P., Picuno, C., Manera, The effect of soil solarization and protection techniques on yield traits of melon in unheated greenhouse, Acta Hortic. 559(2) (2001) 705–712. https://doi.org/10.17660/ActaHortic.2001.559.104

[58] Kyrikou, D. Briassoulis, Biodegradation of agricultural plastic films: A critical review, J. Polym. Environ. 15(2) (2007) 125–150. https://doi.org/10.1007/s10924-007-0053-8

[59] N. Lucas, C. Bienaime, C. Belloy, M. Queneudec, F. Silvestre, J.E. Nava-Saucedo, Polymer biodegradation: Mechanisms and estimation techniques, Chemosphere. 73 (2008) 429–442. https://doi.org/10.1016/j.chemosphere.2008.06.064

[60] K.G. McCabe, J.A. Schrader, S. Madbouly, D. Grewell, W.R. Graves, Evaluation of biopolymer-coated fiber containers for container-grown plants, Hort. Technology 24 (2014) 439-448. https://doi.org/10.21273/HORTTECH.24.4.439

[61] M. Malinconico, B Immirzi, S. Massenti, F.P. LaMantia, P. Mormile, L. Petti, Blends of polyvinylalcohol and functionalized polycaprolactone. A study of the melt extrusion and post-cure of films suitable for protected cultivation, J. Mater. Sci. 37 (2002) 4973–4978. https://doi.org/10.1023/A:1021058810774

[62] M. Avella, E. DiPace, B. Immirzi, G. Impallomeni, M. Malinconico, G. Santagata, Addition of glycerol plasticizer to sea weeds derived alginates: Influence of microstructure on chemical–physical properties, Carbohyd. Polym. 69 (2007) 503–511. https://doi.org/10.1016/j.carbpol.2007.01.011

[63] Simkovic, What could be greener than composites made from polysaccharides?, Carbohyd. Polym. 74 (2008) 759–762. https://doi.org/10.1016/j.carbpol.2008.07.009

[64] A.K. Mohanty, M. Misra, G. Hinrichsen, Biofibres, biodegradable polymers and biocomposites: An overview, Macromol. Mater. Eng. 276/277 (2000) 1–24. https://doi.org/10.1002/(SICI)1439-2054(20000301)276:1<1::AID-MAME1>3.0.CO;2-W

[65] H. Hatami-Marbini, S. Pietruszczak, On inception of cracking in composite materials with brittle matrix, Comput Struct. 85 (2007) 1177–1184. https://doi.org/10.1016/j.compstruc.2006.12.001

[66] L.M. Lewin, E.M. Pearce, Handbook of fiber science and technology, Fiber chemistry, vol. IV, Marcel Dekker, New York, 1985.

[67] P. Wambua, U. Ivens, I. Verpoest, Natural fibers: can they replace glass in fiber-reinforced plastics?, Composite. Sci. Technol. 63 (2003) 1259–1264. https://doi.org/10.1016/S0266-3538(03)00096-4

[68] V.P. Cyras, J. Martucci, S. Iannace, A. Vázquez, Influence of the fiber content and the processing conditions on the flexural creep behavior of sisal–PCL–starch composites, J. Thermoplast Compos. 14 (2001) 1–13.

[69] D.G. Hepworth, D.M. Bruce, The mechanical properties of a composite manufactured from non-fibrous vegetable tissue and PVA, Compos Appl Sci Manuf. 31 (2000) 283–285. https://doi.org/10.1016/S1359-835X(99)00100-1

[70] J.A. Schrader, G. Srinivasan, D. Grewell, K.G. McCabe, W.R. Graves, Fertilizer effects of soy-plastic containers during crop production and transplant establishment. Hort Science. 48 (2013) 724-773. https://doi.org/10.21273/HORTSCI.48.6.724

[71] S. Sahoo, A. Behera, R.M. Nanda, R. Sahoo, P.L. Nayak, Gelatin blended with Cloisite 30B (MMT) for control release of Ofloxacin, Am J Sci Ind Res. 2(3) (2011) 363-368. https://doi.org/10.5251/ajsir.2011.2.3.363.368

[72] M.S. Helgeson, Horticultural evaluation of zein-based bioplastic containers, M Sc. Thesis, Iowa State University, United States, 2009.

[73] Sun, H. Huang, F. Sun, G. Wu, Z. Chang, Degradable Nursery Containers Made of Rice Husk and Cornstarch Composites, Bio. Resources. 12(1) (2017) 785-798. https://doi.org/10.15376/biores.12.1.785-798

[74] Poggio, E. Ciannamea, L. Castillo, S. Barbosa, Desarrollo de recipientes activos y biodegradables para cultivos agrícolas, Avances en Ciencias e Ingeniería. 7(2) (2016).

[75] Van de Wetering, S. Athalage, Biodegradable planters, U.S. Patent No. 7681359, 2010.

[76] Cameron, P. Styles, Biodegradable plant pots, U.S. Patent Application No. 12/098666, 2009.

[77] J.F. Whitehead, Biodegradable plant pot, U.S. Patent No. 8474181, 2009.

[78] S. Hermann, Planting container and method of making the container, U.S. Patent No. 6092331, 2000.

[79] P. Kelly, A. Lynch, Plant pots, plant and tree guards, and plant and tree wrap, WIPO Patent, 2007.

[80] W. Waldenmeier, Biodegradable plant pot-formed from flat sheet e.g. corrugated cardboard with polyester coating. Germany Patent, 1991.

[81] Haimin, Nutritive flowerpot prepared from pig manure, China Patent No. 107736150, 2018.

[82] M. Sudo, S. Ueda, K. Yagi, Nursery container and molding of the same. Japan Patent No. 10309135, 1998.

[83] S. Qingxi, H. Jianxiang, Plant fibre pot for seedling and method for mfg. same. China Patent No. 1268283, 2000.

[84] C. Samet, Bio-degradable compositions and use thereof. U.S. Patent Application No 15/959633, 2018.

[85] Wenli, Degradable flowerpot. China Patent No. 201393428, 2010.

[86] S. Cheung, H. Liang, Pot de fleur vert naturellement degradable et son procede de fabrication. WIPO Patent No. 2007051393, 2007.

[87] Gibert, Biodegradable pot or bucket, France Patent No. 3051097, 2017.

[88] R. Conneway, S. Verlinden, A.K. Koeser, M. Evans, R. Schnelle, V. Anderson, J.R. Stewart, Use of alternative containers for long- and short-term greenhouse crop production, Hort. Technology. 25(1) (2015) 26–34. https://doi.org/10.21273/HORTTECH.25.1.26

[89] Y. Sun, G. Niu, A.K., Koeser, G. Bi, V. Anderson, K. Jacobsen, R. Conneway, S. Verlinden, R. Stewart, T. Sarah, S.T. Lovell, Impact of biocontainers on plant performance and container decomposition in the landscape, Hort. Technology. 25(1) (2015) 63-70. https://doi.org/10.21273/HORTTECH.25.1.63

Chapter 6

Biodegradable Packing for Non-Food Items

Debanga Bhusan Konwar, Titash Mondal*, Shreedhar Bhat

Corporate R&D Center, Momentive Performance Materials Inc., Survey No. 09, Hosur Road, Electronic City (West), Bangalore 560100, India

*titash.mondal@momentive.com

Abstract

In the current perspective, the use of packaging material for applications ranging from food items to personal care applications is on the rise. However, most of the packaging materials currently used are non-biodegradable in the environment. Hence, development of environmentally friendly packaging material is critical for maintaining a proper ecosystem of the world. In the current chapter, we reflect on the different types and classification of packaging material used in the market, with special emphasis on non-food items. A comprehensive overview is provided on the synthetic perspective of the biodegradable polymers used for packaging. Holistically, how biodegradable polymers are leveraged by different industries for developing non-food packaging is also discussed in this chapter.

Keywords

Biodegradable Packaging, Electronics Packaging, Personal Care Packaging, Horticulture, Dunnage

Contents

1.   Introduction

Packaging material can be industrially classified under two different classes, namely rigid packaging and the flexible packaging. The classification of the packaging material is done based on the ability of the material to deform under stress. Thus, it can be reasonably inferred that the selection of the material is critical in controlling the type of end material to be used for packaging. Common examples of rigid packaging include thermoforms, cans, etc. As the name indicates, the material selection for rigid packaging involves judicious choice of candidates which demonstrates significant impact resistance. Flexible packaging material involves material which can deform under the application of stress. Compared to the rigid packaging, flexible packings are light weight with the caveat that the latter is less durable compared to the former. Common examples of material used for flexible packaging are polyolefin based. Based on the report from Wood Mackenzie, globally polyethylene constitutes 56%, polypropylene 22%, polyethylene terephthalate (10%), polyvinylchloride (1%), nylon (1%) and other polymer (10%) of the market share [1]. Despite the presence of wonderful properties like excellent shelf life, durability, etc., a ubiquitous resistance towards its usage is noted globally. In the current scenario, due to implementation of several policies by governmental organizations, there has been an exponential growth in the sentiment for anti-polyolefin-based packaging material. It is worthy of mentioning, based on the information from multiple agencies, around 40% of these plastics finds its end application consumer goods with a typical usage cycle of one time [2]. However, lack of proper recycling infrastructure and capabilities makes these packaging materials environmentally unfriendly. Eriksen et al. did a listing on total number of floating particles and their floating weight through a survey carried out between the years of 2007-2013 [3]. Based on their report, the North Pacific Ocean was the most polluted followed by Indian Ocean and North Atlantic

Ocean. Jambeck et al. reported an alarming number of 6-12 million tons of plastics that enters the oceans [4]. Such a situation is unwanted in terms of maintaining a healthy marine ecosystem.

This gave an impetus to the researchers from different parts of the globe to pursue research in the field of biodegradable polymers for application as a sustainable packaging material. Biodegradable polymers have been considered as the solution to the problems associated with the non-biodegradable counterparts.

2.   Biodegradable polymers

The following section of the chapter will provide insight to different biodegradable polymers, their mechanical stability and their synthesis perspective.

According to IUPAC definition, biodegradable polymers are those polymers which are susceptible to degradation with lowering of molar mass by biological activity e.g. catalytic activity of enzymes or hydrolysis [5]. Biodegradable polymers can be divided as naturally occurring, synthetic and hybrid biodegradable polymers as shown in Fig.1. The naturally occurring polymers include starch, cellulose, chitin, pectin, alginic acid, natural gums, and lignin which can be extracted from tree-based sources. Additionally, materials from animal sources like chitosan, gelatin and collagen are also classified as natural source of polymer. Examples of synthetic polymers from natural resources include polylactide, polyhydroxybutarate while polycaprolactone is a representative example of biodegradable polymer from non-renewable resources.

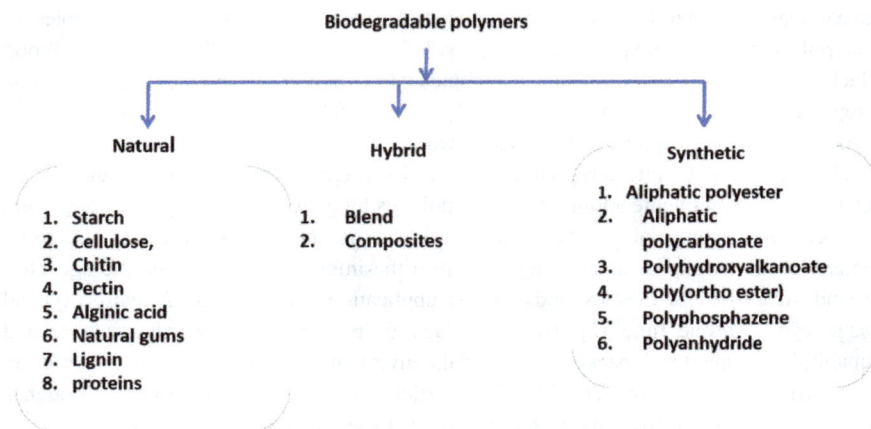

Biodegradable polymers

Natural	Hybrid	Synthetic
1. Starch	1. Blend	1. Aliphatic polyester
2. Cellulose,	2. Composites	2. Aliphatic polycarbonate
3. Chitin		3. Polyhydroxyalkanoate
4. Pectin		4. Poly(ortho ester)
5. Alginic acid		5. Polyphosphazene
6. Natural gums		6. Polyanhydride
7. Lignin		
8. proteins		

Figure 1. Classification of biodegradable polymers

Poly (lactic acid) (PLA) is the most widely used biodegradable polymer due to its good mechanical strength, ease of process ability, high melting temperature, and biocompatibility along with biocompatible in nature [6]. Additionally, PLA can be manufactured from various bio-based resources such as trapioca, corn starch and sugar cane. In additions, having a high mechanical strength along with high melting point, makes PLA a perfect candidate in many high demand applications [7]. Also, PLA has good process ability and brings various attributes to packaging applications, including strength, transparency, twist retention, low-temperature heat seal ability along with excellent barrier resistance. Additionally, lactic acid which is precursor of PLA, exists in D and L forms and PLA behaves as amorphous to semi crystalline polymer depending upon L and D content. All those combined advantages make PLA the most promising and attractive of sustainable polymers for various non-food packaging applications [8].

Poly (glycolic acid) PGA is another semi-crystalline biodegradable polymer produced by ring opening of glycolide or polycondensation reaction of glycolic acid. PGA degrades in 2 to 4 weeks, losing 60% of its mass during the first two weeks and is use mostly in biomedical applications. However, it has potential to be use in packaging applications due to its high degree of crystallization which results in exceptional barrier properties. It can be used in different films e.g. multilayer stretch blow-molding, biaxially oriented films, and blown films which are precursors for many packaging applications. Additionally, PGA can be blended with different conventional polymers such as PP, PE, PET and polyamide which gives an added advantage to its use in various food packaging as well as in non-food packaging applications. It can also be easily modified by copolymerization and blending with PLA to form a range of material with different properties which are already in use in various biomedical applications [9].

Polycaprolactone (PCL) is one of the highly explored petroleum sourced another synthetic biodegradable polymer. PCL is a semi-crystalline hydrophobic polymer with melting point of 59–64 °C and its crystallinity tends to decrease with increasing chain length. PCL shows blend-compatibility with many biodegradable and non-biodegradable polymers and has explored into its potential application in various fields. PCL can be synthesized by the ring-opening polymerization of the cyclic monomer caprolactone. PCL shows elongation at break (%) of ~ 500 and hence can be blended with other polymers to improve toughness of any brittle polymers. Depending upon the molecular weight, virgin PCL can have degradation of 2–4 years. The degradation time can be manipulated by copolymerization with other cyclic monomers. Although PCL is mainly used in the biomedical field, some other application e.g. fully biodegradable grocery bags; sutures, fibers, and biodegradable film can be made from PCL [10].

Poly (butylene succinate) (PBS) is another commercially available biodegradable aliphatic polyester with a good melting process ability and thermal and chemical resistance. It is a white semi crystalline polymer with glass transition temperatures of - 45 °C to – 10°C and melting point of 90°C- 120°C. Due to the highly processible nature of PBS, it can be molded into various promising applications e.g. biodegradable packaging film, mulch film, cutlery, containers, bags etc. [11].

Polyhydroxyalkanoates (PHAs) are biobased biodegradable polymers with hard crystalline to elastic properties, depending on the monomer content. PHB shows mechanical properties like Young's modulus and tensile strength of as comparable as PP though elongation at break is lower than that of PP. PHA type polymers can be injection molded and can be use in packaging application [12].

Aliphatic polycarbonates e.g. poly (propylene carbonate), poly (butylene carbonate), poly (hexamethylene carbonate) etc. are also important biodegradable polymers and have the potential to be use in various green technology applications. They can be prepared either by ring opening polymerization of cyclic monomer or polycondensation reaction of dimethyl carbonate and various diols. The application of aliphatic polycarbonatesis motivated by the mechanical resistance, UV resistance, chemical resistance and stability, anti-scratch and anti-stain properties along with biodegradable properties. Also, aliphatic polycarbonates upon degradation do not generate acidic microenvironment unlike polyesters which makes it one of the ideal candidates to use in biomedical applications [13].

2.1   Synthetic perspective of biodegradable materials

Synthetic biodegradable polymers can be mostly synthesized by step-growth polymerization and ring opening polymerization e.g. most widely used synthetic biodegradable polymers aliphatic polyester can be synthesized by both the polymerizations. Another biodegradable inorganic polymer, polyphosphazene can be synthesized via ring opening polymerization. Biodegradable polymers PBS can be synthesized by step growth polymerization using suitable catalyst. In below sections brief reviews of step growth polymerization and ring opening polymerization are made [14].

2.1.1 Step-Growth Polymerization

Step-growth polymerization or polycondensation is a type of polymerization where multifunctional monomers especially bifunctional react to form high molecular weight polymers. Step-growth polymerization is the traditional synthetic route to synthesize many widely used polymers, more specifically biodegradable polymers e.g., aliphatic polyester, polycarbonate, PBS are synthesized by polycondensation reaction. Aliphatic polyester can be synthesized by the reaction between hydroxyl and carboxyl groups. PBS can be synthesized by the reaction between succinic acid and 1,4-butanediol. On the other hand, the polycondensation reaction between hydroxyl group and carbonate produce an aliphatic polycarbonate.

In a step-growth polymerization, the molecular weight of the polymer chain builds up slowly. Monomers react to form first dimers, then trimers, longer oligomers and eventually long chain and to get polymer with high molecular weights, the conversions should be very high (>98-99%). Carothers equation (Equation 1) proves that polymers with high molecular weight only can be achieved at high conversion.

$$DP = \frac{1}{1-P} \tag{1}$$

Where, DP is the degree of polymerization and p is the extent of reaction.

To get high molecular weight polymers by step growth polymerization, some special attention needs to be given in the reaction conditions. First, a stoichiometric balance of the two difunctional monomers is very important to get high molecular weight polymers. Stochiometric imbalances lead to end capping of oligomers which terminated the polymerization results in poor molecular weight polymer. Secondly, monofunctional impurity or any other impurity should be completely absent to achieve high molecular weight in step growth polymerization reaction. The presence of impurity can act as caps to the oligomer chain which terminates the polymerization. Third, the step growth polymerization should be a very high yield reaction without any side reactions. Additionally, use of high vacuum, high temperature or absorbents such as molecular sieves to remove the byproducts required to achieve high molecular weight polymers in the step growth polymerization method.

In general, step-growth polymerization follows a two-step procedure; first oligomers are formed and then in the second step further condensates oligomers to get high molecular weight polymers [15].

Oligomer formation or esterification (in case of synthesis of polyester) is the first step of step-growth polymerization where low molecular weight oligomers are formed due to a low equilibrium constant. This step usually carried out in moderate temperature and without application of a vacuum. Here at the completion of the step, theoretical amount of byproduct e.g. water and methanol can be collected from the reaction mixture. In this step, oligomers (dimer, trimers, tetramers) are formed which can be monitored by NMR technique.

Advanced Applications of Bio-degradable Green Composites   Materials Research Forum LLC
Materials Research Foundations **68** (2020) 138-155   https://doi.org/10.21741/9781644900659-6

In the polycondensation step, oligomers prepared in the first step eventually undergo polymerization to get high molecular weight polymers. Here, the reaction is carried out at a high temperature with the application of a vacuum to achieve high molecular weight polymers. The reaction time and temperature depend upon the reacting monomers and their thermal stability of the resultant polymer. As the reaction progresses viscosity of the reaction mixture increases and can be monitored.

Step-growth polymerization based different synthetic routes for some synthetic biodegradable polymers are shown in Scheme 1.

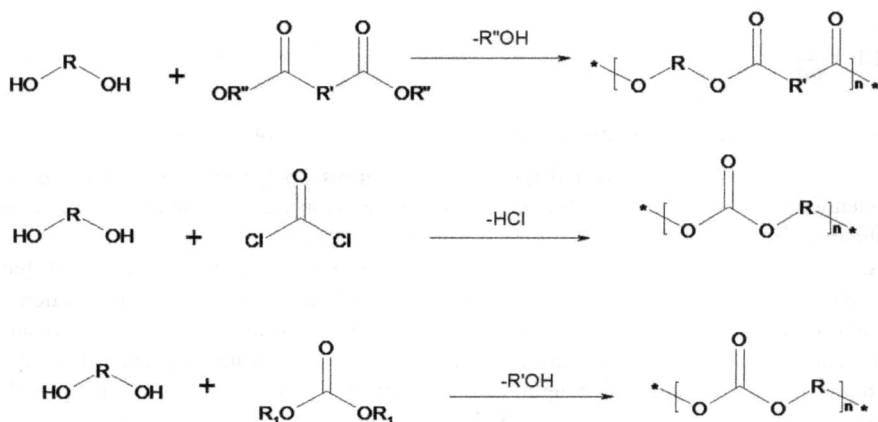

Scheme 1: Synthesis scheme of aliphatic polyester and aliphatic polycarbonate using step-growth polymerization

2.1.2   Ring opening polymerization (ROP)

Ring-opening polymerization is another important method to successfully synthesize various biodegradable polymers and is used for the commercial production of variety of biodegradable polymers. The ROP yields polymers with higher molecular weight with narrow dispersity in relatively shorter time which is normally difficult to achieve from step-growth polymerization [16]. It does not require precise stoichiometric calculation of functional groups which is very an important aspect for the synthesis of high molecular weight polymer in the step polycondensation reaction. ROP can be used to make

Advanced Applications of Bio-degradable Green Composites   Materials Research Forum LLC
Materials Research Foundations **68** (2020) 138-155   https://doi.org/10.21741/9781644900659-6

copolymers with specific properties for desired application such as packaging, coatings, fibers, elastomers, adhesives, and composites. Additionally, ROP is proceeds via living manner hence it gives control over the chemistry of polymerization accurately and thus, the properties of resultant polymers, such as molecular weight, and architecture can be varied as per the requirements. Also, due to the living polymerization nature of ROP, the synthesized polymer can be modified with reacting other functional group to produce different block copolymers [16].

The driving force behind the ROP is the relief of ring strain of cyclic monomer. Cyclic monomers are monomers with ring structure in which one or more series of atoms are connected. In ROP, the ring gets opened and attached to the chain at either ends promoting chain growth. The four or seven membered rings have a greater thermodynamic driving force to undergo ROP due to their greater ring strain as compared to the five or six membered rings. Moreover, the substituents on the monomer ring has decreased the ring strain hence decrease the driving force to get polymers from these rings. This is attributed to interactions between substituent's is more pronounced in the linear versus the cyclic molecules.

Some biodegradable polymers synthesized via ROP with their monomer are given in the following Scheme 2.

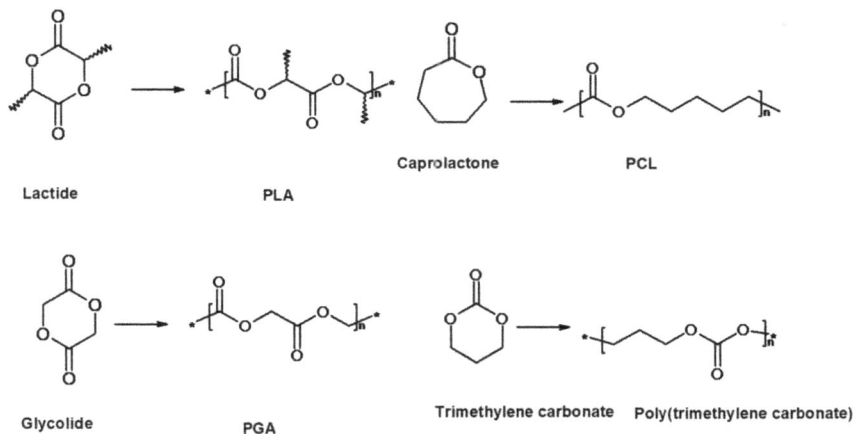

Scheme 2: Structure of some monomer and their corresponding biodegradable polymers by ROP.

Advanced Applications of Bio-degradable Green Composites   Materials Research Forum LLC
Materials Research Foundations **68** (2020) 138-155   https://doi.org/10.21741/9781644900659-6

2.2   Catalysts for ring opening polymerization

Various organometallic derivatives of metals catalyst with are used as in ROP along with enzymatic catalyst. Stannous (II) 2-ethylhexanoate [Sn (Oct)2] is one of the most widely used catalyst system in ROP especially for the synthesis of aliphatic polyester. It gives high yield with high molecular weights with narrow dispersity and is approved by FDA as a food stabilizer. Usually ROP reaction with SnOct2 is carried out in the presence of active hydrogen compounds e.g. alcohols, amine etc. or may be from impurities if present. Kinetics and MALDI–MS studies prove that ROP by Sn (Oct)2 proceeds through coordination-insertion mechanism. The following steps are involved in the ROP by Sn (Oct)2 (Scheme 3):1. The weak complexation is formed by monomer to tin alkoxide. 2. Coordination insertion process takes place by coordination of Lewis acid metal with ester carbonyl carbon of lactide. 3. Alkoxide moiety attacks the carbonyl carbon forming a tetrahedral intermediate which collapse by acyl carbon-OR bond to open the ring of the monomer [17].

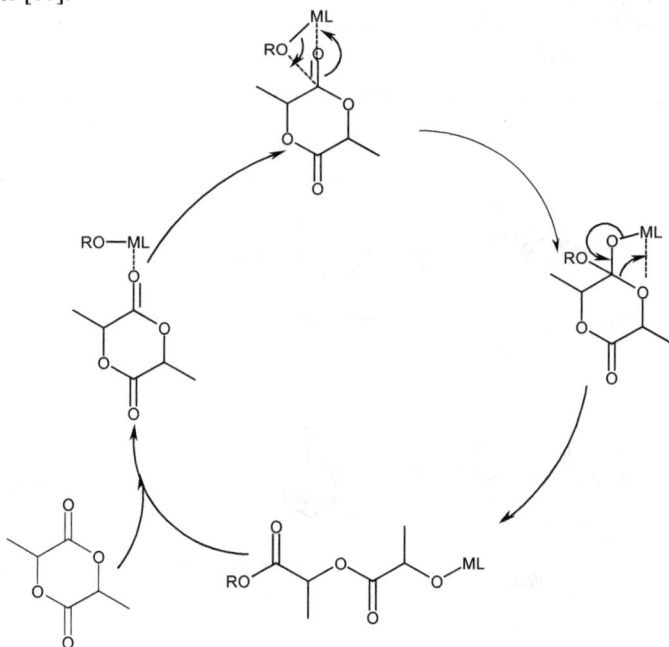

Scheme 3 Coordination insertion mechanism of ROP of lactide [17]

3.   Biodegradable polymers for non-food item packaging

Naturally occurring polymers often find limited usage in packaging industries due to their poor physical properties. For instance, from the molecular perspective, the hydroxyl side chain dangling from the cellulosic structure imparts greater extent of crystallinity to the material. As a result, the final material tends to demonstrate poor mechanical properties. Even though neat cellulosic material demonstrates poor properties, however, nanowhiskers generated from cellulose are a potential candidate for packaging industries [18]. Additionally, efforts are made to improve the properties of naturally occurring biodegradable polymers and used in many applications e.g., polylactide/lignin composites are used in many short-term ecofriendly applications such as much films and grocery bags etc. There are also some naturally occurring biodegradable polymers starch based blends products introduced by Novament, Biotec with promising properties. However, the synthetic biodegradable polymers e.g. aliphatic polyesters and aliphatic polycarbonate and their blends and their composites are extensively used in many industrial applications [19, 20].

Some conventional synthetic biodegradable polymers possess a good mechanical performance depending on the molecular weight and as well stereochemical composition which is comparable to many widely used conventional polymers like polystyrene (PS), polypropylene (PP) and poly (ethyleneterephthalate) (PET). A comparative analysis of different properties of biodegradable polymers is given in Table 1.

Table 1. Properties of different biodegradable polymers along with some conventional polymers

Polymer	Tensile strength (MPa)	Youngs Modulus (GPa)	Elongation at-break (%)
Poly(lactic acid)	53	3.4	5.5
Poly(glycolic acid)	68	6.9	15-20
Poly(caprolactone)	20.7	0.21	500
Polystyrene	45	2.9	7
poly (ethyelene terepthalate)	54	2.8	130
Polypropylene	31	0.9	120

3.1 Biodegradable packaging in the electronics industry

In US5177660A patent application, Kilner demonstrates about the invention of a biodegradable and recyclable bag for enclosing electronic circuit for prevention against electrostatic damage. Kraft paper was used for the purpose. The electromagnetic shielding properties were achieved by printing conductive grids over the kraft paper. The current packaging material acts as a potential alternative for metallic enclosures used for such applications. This is a neat initiative towards reducing electronic waste [21].

On a similar note, Chase et al. in their patent application US20090096703A1 reports about the invention of paper/biodegradable plastic laminate for shielding electromagnetic interferences. The shielding properties are imbibed into the matrix by the introduction of a conducting layer between the paper web and biodegradable plastic. The developed composition also claimed to provide electromagnetic shielding and a radiofrequency identification chip [22].

By far it can be noted that biodegradable polymers find application in electronic space. On a similar note, Bradford in his patent application US5613610A demonstrated the utility of biodegradable polymer as a static dissipative packaging material. The current invention relates to the metallization of cellophane and further the metallized cellophane was adhered to paperboard. The resulting composition is biodegradable in nature [23].

A unique invention is reported in the patent application US20160052692A1, wherein the inventor discusses the method of making a biodegradable packing material demonstrating thermally insulation and cooler properties. The main components of the invention include a corn starch based foam paper, a bioplastic and compostable ink. The invention disclosed in the patent application provides a solution for thermally sensitive shipping materials. Resistance to moisture and reduction in friction against the surrounding environment are the additional advantages reported due to the use of corn starch. The bioplastic film referred herein are potato based [24].

US20160194828A1 reports about the development of biodegradable packaging material for electronic products. To prevent the effect of impact on the electronic material during transportation, Styrofoam, expandable polyethylene (EPE) is utilized. In the particular invention, a replacement of synthetic polymers with biodegradable material was disclosed. Biodegradable fibers from pulp and waste paper were used. The steps involved in making the cushioning packaging material involves a) making a composite slurry of talcum powder, water and fibers; b) mixing the slurry into expandable material and glue. The mixture is thereby subjected to treatment in a molding instrument [25].

3.2   Biodegradable packaging in the personal care and home care industry

In EP0739368B1, the inventor reports about PHA copolymer based biodegradable container targeted for applications in personal care, household and agricultural use. Usually, containers for such end application have low life time (usually less than 12 months). Following which they are discarded. This makes the situation critical in terms of the recyclability or composability of the material. Hence the invention mentioned in the patent application provides a solution to the existing challenges [26].

On a similar note, Stanley et al., in their invention disclosed in patent application US8871319B2 talks about the development of flexible barrier packages for items pertaining to personal care industries. The flexible barrier packaging material involves sealant, an adhesive layer and barrier material. The sealant was selected from PHA based material. Incidentally, ultra-low-density polyethylene extracted from sugarcane was also used for similar purpose. The adhesive material used to tie different segments was selected from polylactic acid. The barrier material was selected from polyglycolic acid having metallic fillers in them [27].

Bauer et al. in their patent application WO2007135037A1 reports about the development of flexible packaging material for moisture sensitive material. Polylactic acid and polyvinyl alcohol were taken as the lead candidate. They were further turn into 5 layers using blown film extruders. The oxygen permeability reported for such a formulation was reported to be 1.6 cc/m2[28].

Iyengar et al. in their patent application, US8771835B2 discloses about the development of high barrier packaging material based on biodegradable resources. The developed material demonstrated an excellent reduction in the moisture vapor transmission rate (Less than 0.3 grams per 100 square inches per day). The barrier layer used for the same involved metallized PLA mostly [29].

Penttinen et al. in their patent US9181010B2 disclosed about the invention of heat sealable packaging material using biodegradable material, wherein PLA was used as the lead molecule. It is worth mentioning that PLA based packaging materials are often associated with problems like, brittleness. Further, in terms of the processing aspect, these PLA based materials require higher temperatures for extrusion. Hence, to circumvent the problems associated with PLA, they are often coextruded with biodegradable co-polyester, cellulose ester and polyester amides. Further, specific attributes like heat sealing ability are often hindered due to the PLA (known for its high melting point). Hence, biodegradable polyesters are co-blended with PLA for developing heat sealable compositions [30].

Wnuk et al. in their patent US8367173B2 reports the development of biodegradable sachet for personal care products like shampoo, conditioner, etc. The uniqueness of the developed composition was its resistance towards transmission of moisture, even at high relative humidity. The developed composition demonstrated a moisture vapor transmission rate of 10 $g/m2/day$ at 37° C and 90% relative humidity. Co-extrusion technique was adopted to develop the multilayer sachet developed in this invention. PHA was used as the biodegradable sealant, while the barrier material was polyglycolic acid based [31].

Suskind in patent US5458933A relates to the invention of biodegradable packaging material mostly for liquid material like shampoo, conditioner, wet wipes, etc. The developed packaging material demonstrates key attributes like heat sealable, flexible, adhesive, resistance to stress cracks and abrasion resistant. Polycaprolactone was selected as the model material [32].

Lignin is one of the most commonly used materials employed by the packaging industry for decades. The unique structure of the lignin with myriads of antioxidant functionalities over its surface makes it an ideal candidate for packaging requiring resistance towards light and oxidation. Domenek et al. reported about the possibilities of using lignin as an active material in the packaging [33].

US20130053293A1 relates to the invention of biodegradable packaging for detergents. The core material for the packaging involves polymers not limited to polylactic acid, polyhydroxyalkanoates, polyhydroxybutyrate, polycaprolactone, polyhydroxyhexanoate. The uniqueness of the current invention is that the developed material is resistant to water and is tolerant for usage up to 70°C [34].

3.3 Biodegradable packaging in horticulture

Almenar et al. in their patent US20100151166A1 reports about the development of micro-perforated packaging material based on polylactic acid. The specific application targets the packaging for flowers. Microperforated films helps in balancing the appropriate ratio of CO_2 to O_2. The added advantage includes improved shelf life of the material packed using the microperforated films. Cold micro needles were used to create the micro perforation of the PLA films. [35]

On a similar note, US20110225882A1 reflects to floral packaging materials from biodegradable polymers. The wrapping material was selected from polylactic acid, cellulose, polyhydroxyalkanoates, and their copolymers. [36]

It is worth mentioning that plastic originated from petroleum sources are often employed as container in the floriculture industry for planting trees. This results in generation of large number of plastic containers, which essentially are non-biodegradable. Hence, emphasis is laid upon developing containers which are biodegradable. The uniqueness of the biodegradable container is that the plant can be directly planted in the soil. Further, the presences of microorganism in the soil break down the container to carbon dioxide, water and biomass [37]. On a similar note, rice straw and starch adhesives modified by polyvinyl alcohol was employed by Wu et al. to make biodegradable container for the floriculture industry [38]

3.4   Biodegradable packaging in dunnage

US4997091A relates to the invention of shock absorbing dunnage packaging. The current invention involves a generation of pulp fibers by recycling of scrap paper in hot water. The pulp fibers thereby obtained were extruded in the form of dunnage package material. [39]

The utility of the scrap paper was utilized for developing loose fill packaging material and is reported in US5900119A. Loose fill packing materials are used to support an article being transported in a container. The cellulose pulp extracted from the scrap paper is subjected to molding in the presence of air breathable, stretchable fabric. Further, a vacuum was applied to the mold to deposit the pulp over the fabric. Such a composite was further utilized for loose fill packaging. [40]

4.   Commercial source of biodegradable polymers for packaging

With the ever-growing demand for reduction in carbon footprint and a focus on sustainability, different industries across the globe have focused on commercializing biodegradable polymers. The following section of the article collates the commercial source of the biodegradable polymers commonly used in the packaging industries (Table 2).

Table 2: Commercial Source of Biodegradable Polymers for Packaging

Polymer	Industry	Commercial Name
Polylactic Acid	BASF	Ecovio
	Naturework	Natureworks PLA
	Stanelco	Starpol 2000
	Euromaster	Bioter PLA
Polycaprolactone	Solvay	Capa PCL/Capa 6500C
Starch Based	Novamont	Mater bi-starch

Concluding remarks

In summary it can be unequivocally suggested that polymers with biodegradable properties could be the most innovative materials being developed in the packaging industry. A ubiquitous resistance towards its usage of plastic material from petroleum sources is noted globally. In the current scenario, due to the implementation of several policies by governmental organizations, there has been an exponential growth in the sentiment for anti-polyolefin-based packaging material. Thus, biodegradable polymers for packaging become a unique choice for the industries. Biodegradable polymers already find wide application in food packaging, however, it can be noted that use of biodegradable polymers as a packaging material beyond the food industry is well established. From the trend in the patent filing discussed above, it can be reasonably inferred that biodegradable packing is an important contributor for industries like personal care and home care, electronics industry, horticulture industries as well as cargo industries.

References

[1]   Polyethylene Terephthalate (PET) Packaging, Wood Mackenzie. https://www.woodmac.com/research/products/chemicalspolymersfibres/polymers/polyethylene-terephthalate-packaging/ (Accessed in 10 July 2019)

[2]   Pressure to Reduce consumption of Single Use Plastic Packaging will continue into 2019 https://www.plasticstoday.com/packaging/pressure-reduce-consumption-single-use-plastic-packaging-will-continue-2019/8501551360001

[3] M. Eriksen, L.C.M. Lebreton, H.S. Carson, M. Thiel, C.J. Moore, J.C. Borerro Plastic pollution in the world's oceans: more than 5 trillion plastic pieces weighing over 250,000 tons afloat at sea, PLoS ONE. 9 (2014) 111913. https://doi.org/10.1371/journal.pone.0111913

[4] J.R. Jambeck, K. Johnsen, Citizen-based litter and marine debris data collection and mapping, Comput. Sci. Eng. 17 (2015) 20–26. https://doi.org/10.1109/MCSE.2015.67

[5]   Compendium of Polymer Terminology and Nomenclature, IUPAC 1139 recommendations, RSC publishing 2008.

[6] L.S. Nair, C.T. Laurencin, Biodegradable polymers as biomaterials, Prog. Polym. Sci. 32 (2007) 762-798. https://doi.org/10.1016/j.progpolymsci.2007.05.017

[7] N.K. Madhavan, N.R. Nair, R.P. John, An overview of the recent developments in polylactide (PLA) research, Biores Tech. 101 (2010) 8493-8501. https://doi.org/10.1016/j.biortech.2010.05.092

[8] R.M. Rasal, A.V. Janorkar, D.E. Hirt, Poly (lactic acid) modifications, Prog. Polym Sci. 35 (2010) 338-356. https://doi.org/10.1016/j.progpolymsci.2009.12.003

[9] Z. Zhen, O. Ophir, G. Ritu, K. Joachim, Biodegradable Polymers, Principles of Tissue Engineering (Fourth Edition). (2014) 441-473. https://doi.org/10.1016/B978-0-12-398358-9.00023-9

[10] A.W. Maria, W.H. Dietmar, The return of forgotten polymers- polycaprolactone in the 21st century, Prog. Poly Sci. 35 (2010) 1217-1256. https://doi.org/10.1016/j.progpolymsci.2010.04.002

[11] R. Ewa, Compostable polymer properties and packaging applications, Plastic Films in Food Packaging. 7 (2013) 217-248. https://doi.org/10.1016/B978-1-4557-3112-1.00013-2

[12] Z . Li, J. Yang, X. Jun Loh, Polyhydroxyalkanoates: Opening doors for a sustainable future, NPG Asia Material. 8 (2016) 265-274. https://doi.org/10.1038/am.2016.48

[13] G. Rokicki, Aliphatic cyclic carbonates and spiroorthocarbonates as monomers, Prog. Polym. Sci.25 (2000) 259-342. https://doi.org/10.1016/S0079-6700(00)00006-X

[14] M. Okada, Chemical syntheses of biodegradable polymers, Prog. Polym. Sci. 27 (2002) 87-133. https://doi.org/10.1016/S0079-6700(01)00039-9

[15] D.N. Bikiaras, D.S. Achilia synthesis of poly(alkylene succinate) biodegradable polyesters, part ii: Mathematical modelling of the polycondensation reaction, Polymer. 49 (2008) 3677-3685. https://doi.org/10.1016/j.polymer.2008.06.026

[16] O. Dechy-Cabaret, B. Martin-Vaca, D. Bourissou, Controlled ring-opening polymerization of lactide and glycolide, Chem. Rev. 104 (2004) 6147-6176. https://doi.org/10.1021/cr040002s

[17] Y. Zhu, C. Romain, C.K. Williams, Sustainable polymers from renewable resources, Nature. 540 (2016) 354-362. https://doi.org/10.1038/nature21001

[18] J.E. Stephen, Cellulose nanowhiskers Promising materials for advanced applications, Soft Matter. 7 (2011) 303-315. https://doi.org/10.1039/C0SM00142B

[19] Y.L. Chung, J.V. Olsson, R.J. Li, C. W. Frank, R.M. Waymouth, S.L. Billington, E. S. Sattely, Renewable lignin-pla copolymer and application in biobased composites, ACS Sustainable Chem. Eng. 1 (2013) 1231−1238. https://doi.org/10.1021/sc4000835

[20] M. Vert, Aliphatic polyesters: great degradable polymers that cannot do everything. Biomacromolecule. 6 (2005) 538-546. https://doi.org/10.1021/bm0494702

[21] E.K. George , Biodegradable and recyclable electrostatically shielded packaging for electronic devices and media, US Patent US5177660A (1993).

[22] C. Adam, E. James, J. Bowden, Paper/biodegradable plastic laminate and electromagnetic shielding material, US Patent US20090096703A1 (2009).

[23] A.J. Bradford, Naturally degradable and recyclable static-dissipative packaging material, US Patent US5613610A (1997)

[24] B. James, Biodegradable packaging for shipping, US Patent US20160052692A1 (2014).

[25] C. Kun-Hsiang , Method for manufacturing environmentally friendly cushioning material, US Patent US20160194828A1 (2016).

[26] N. Isao Biodegradable copolymers and plastic articles comprising biodegradable copolymers, European Patent EP0739368B1 (2003).

[27] K. Scott, N. Stanley, B. Scott, J. Andrew, J. Wnuk, C. Hayes, E. Charlotte, B.Lee Arent flexible barrier packaging derived from renewable resources, US Patent US8871319B2 (2014).

[28] M. Bauer, K. Mauser, R. Kelm, K. Stark Method for the production of a biodegradable plastic film, and film PCT WO2007135037A1 (2008).

[29] I. Gopal, B. Thomas, L. Gerald, Substantially biodegradable and compostable high-barrier packaging material and methods for production, US Patent US8771835B2 (2014).

[30] P. Tapani, N. Kimmo, K. Tapio, K. Sami heat-sealable biodegradable packaging material, a method for its manufacture, and a product package made from the material, US Patent US9181010B2 (2015).

[31] J. Andrew, S. Wnuk, S. Kendyl , M. John, E. Layman Robert, M. Emily Charlotte, B. Lee, A. Mathew, Degradable sachets for developing markets, US Patent US8367173B2 (2012)

[32] P. Stuart, Suskind compostable packaging for containment of liquids, US Patent US5458933A (1995).

[33] D. Andra, L. Abderrahim, G. Alain, B. Stéphanie, Potential of lignins as antioxidant additive in active biodegradable packaging materials, Journal of Poly. Env. 21 (2013) 692–701. https://doi.org/10.1007/s10924-013-0570-6

[34] D. John, D. Elise, F. Brian, J. Marck, Biodegradable package for detergent, US Patent US20130053293A1 (2010).

[35] A. Eva, A. Rafael A. Hayati, S. Bruce, R. HarteMaria, Micro-perforated poly(lactic) acid packaging systems and method of preparation thereof, US Patent US20100151166A1 (2010).

[36] E. Donald, Weder floral packaging formed of renewable or biodegradable polymer materials US Patent US20110225882A1 (2011).

[37] H.Y. Huang, G.F. Wu, E.H. Sun, Z.Z. Chang, The Influence of heat treatment on the properties of breeding bio-container, Appl. Mech. Mater. 341 (2013) 119-123. https://doi.org/10.4028/www.scientific.net/AMM.341-342.119

[38] G. Wu, E. Sun, H. Huang, Z. Chang, Y. Xu, preparation and properties of biodegradable planting containers made with straw and starch adhesive, Bio. Res. 8 (2013) 5358-5368. https://doi.org/10.15376/biores.8.4.5358-5368

[39] S.M. James, Package containing biodegradable dunnage material, US Patent US4997091A (1999)

[40] J.L. Goers, S. Thomas, H.O. Warda, H.O. William, Method of forming improved loose fill packing material from recycled paper, US Patent US5900119A (1999).

Advanced Applications of Bio-degradable Green Composites
Materials Research Foundations **68** (2020) 156-195

Materials Research Forum LLC
https://doi.org/10.21741/9781644900659-7

Chapter 7

Application of Biodegradable Lipid Composites in Drug Delivery

Vishal Beldar[1], Ritika Joshi[1], Sujit Kumar Ghosh[2], and Manojkumar Jadhao[1]*

[1]Institute of Chemical Technology Mumbai, Marathwada Campus, Jalna, Maharashtra 431 203, India

[2]Department of Chemistry, Visvesvaraya National Institute of Technology, Nagpur, Maharashtra 440 010, India

manojjadhao04@gmail.com, mm.jadhao@marj.ictmumbai.edu.in

Abstract

With the advent of nanotechnology, considerable advancements in the realm of drug delivery have been witnessed with respect to improvements in drug efficacy and safety. Among the various novel drug delivery systems, lipid-based excipients have demonstrated clinical benefit till date. However, parameters like aqueous solubility, permeability, stability, and targeting, in addition to the design of sustainable lipid based drug delivery platforms, still pose a challenge for the researchers. The most recent developments, using lipid composites, have been focused on these issues. Hence this chapter highlights the myriad of lipid composites implemented as drug delivery carriers against lethal diseases and unfolds the promising future of these platforms in nanomedicine research.

Keywords

Drug Delivery, Lipids, Liposomes, Nanomaterials, Lipid Composites

Contents

1.   Introduction

The chemically engineered drug delivery systems (DDS) have tremendously impacted on the treatment methodology of various lethal diseases because of the crucial role they play in controlling the dosing rate of the therapeutic agent to the target site, minimizing their side effects, and extending patient's life span. [1–4] DDS act as drug reservoirs which not mere improve the pharmacological activity but also therapeutic properties of drugs. There are many drugs which have been discovered as a consequence of advanced molecular biology research and that are reasonably effective to treat a disease, but have failed in delivery issues due to biological barriers such as blood brain barrier, blood-eye barrier, the small intestine, nasal, skin and the mouth mucosa, etc. Nature has designed these obstacles to keep unwanted material at bay and allow only small molecules with specific characteristics to permeate through.   Many a times, drugs may have unacceptable secondary effects due to the drug interacting with non-targeted healthy tissues. For many diseases and infections, the pharmaceutical industry can restrict itself to small molecules only but in other cases it has to use larger molecules or biomolecules as the next generation of drugs. However, delivery of large molecules is quite difficult comparted smaller ones thereby building up the challenges associated with drug delivery of larger molecules. Meager aqueous solubility and low dissolution rates of drugs are amongst the prominent challenges faced by the researchers with regard to their bioavailability.[5–7]

There are several different drug delivery systems developed or being developed continuously across the world to make drug delivery systems more accurate and precise. One of such drug delivery vehicle which has occupied a hot-spot and is the preferred choice of the scientific community is the lipid-based drug delivery systems (LBDDS), as it possesses remarkable characteristics such as its biocompatibility, biodegradability, encapsulation capacity for both small and large molecules as well as controlled delivery of drugs. [8] LBDDS can be modified in various ways to meet the different requirements as per the pathological condition, toxicity, drug aggregation, drug encapsulation capacity, side effects of the drug and routes of administrations such as oral, topical, dermal/transdermal, and vaginal, pulmonary or parenteral delivery. [9–13] LBDDS may also enhance the drug potency by reducing the biological degradation or transformation of the active drug. Although, the conventional lipid material has proven its mettle as a potential drug delivery system with many striking features, some of the major drawbacks for the liposomes use are lack of target specificity, inability to perform sustained drug delivery for a longer time period, and rapid degradation. Nevertheless, as the day passes, understanding the complexity and sources of many diseases such as cancer, alzheimer, heart disease, diabetes, malaria, etc. demands a need for a smarter delivery system to reduce the side effects and increase the efficiency of the drugs. Therefore, new generation of LBDDS has gained prominence in recent years commonly termed as lipid composites. Lipid composites are the complex systems composed of extra constituent in the form of gel (hydrogel), polymer (biodegradable as well as high and low density polymer), nanomaterial or biomolecules which may overcome the lacunae of conventional LBDDS. For instance, strategies such as surface modification of liposomes with polymer such as polyethylene glycol (PEG) or incorporation of the pre-encapsulated drug-loaded liposomes inside the polymer-based systems can be implemented to control the release of the therapeutic payloads. Taken together, these approaches can be used to overcome the limitation of materials, liposome and polymer in a way that the instability, short half-life, and rapid clearance of liposomes is improved while biocompatibility index of polymer is increased due to presence of liposome. [14,15] Recently, surface engineered liposome which are functionalized with peptide, protein, and antibody are more in the spotlight due their higher specificity towards the target. [16–18] On these premises, this chapter will focus on a new class of LBDDS which in combination with different materials, can overcome the lacuna in the existing drug vehicle systems. This new generation of LBDDS is mutually referred to as lipid composites.

2.   Why lipids as drug delivery tool

Lipids are naturally occurring organic substances which are relatively insoluble in water but soluble in organic solvents (alcohol, ether etc.). Lipids (Greek: lipos–fat) are of great significance to the human being. They are the chief constituents for the storage of energy and are instrumental in maintaining the cellular structure and various other biochemical functions. Lipids are broadly classified into hydrolysable (undergoes hydrolytic cleavage) and non-hydrolysable. The hydrolysable lipids are further categorized into simple, complex, derived and miscellaneous lipids (Figure 1). Hydrolysable phospholipids are commonly used for the application of drug delivery.

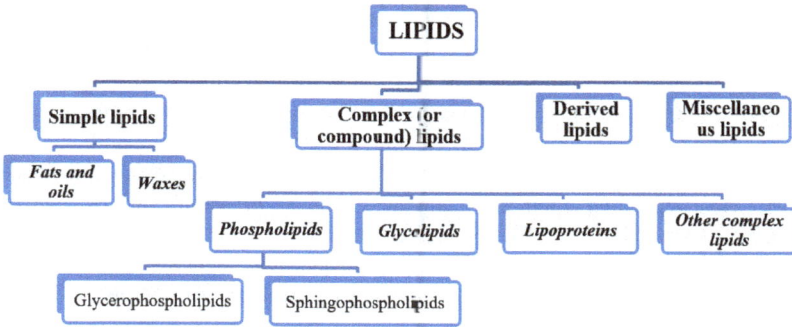

Figure 1: Classifiaction of Lipids.

When these lipids are immersed in aqueous solution above its critical aggregation concentration and above its gel-liquid crystal transition temperature, it forms multilamellar (MLV) or multivesicular vesicles (MVV) having size more than 100 nm. In MLV all the lipid bilayers are concentric whereas in MVV, single larger vesicle encompasses several randomly sized small vesicles. The size of these vesicles can be altered into small and large size unilamellar vesicles (SUV or LUV) (Figure 2) using simple techniques such as ultra-sonication and extrusion. Versatile properties such as enhancing the bioavailability of drugs, reducing toxicity, imparting biodegradability and high encapsulation capacity secure the candidature of lipids for drug delivery systems.

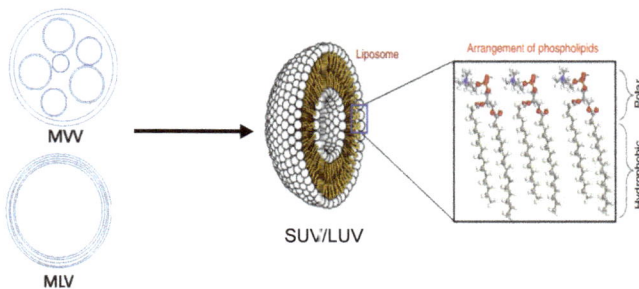

Figure 2: Popular schematic view of MVV, MLV and conventional liposome formed from lipid.[19]

4.1   Bioavailability

The amount as well as the rate of the drug/active moiety with which it reaches the systemic circulation or blood stream in intact form and produce the desirable pharmacological effect is known as the bioavailability. There are various factors affecting the bioavailability of the drug such as its metabolism and solubility. For the enhancement of the bioavailability various approaches have come forth. Among them, use of lipids or LBDDS is the most reliable approach; wherein the principle objective is to enhance the solubility and bioavailability of hydrophobic drugs; reduce the toxicity; increase the efficacy, etc. Depending upon the nature and type of the drug, various type of the application of the lipids and LBDDS has been under investigation. [20] Ahmed and co-workers designed the lipid based system to enhance the brain bioavailability of the antidepressant drug i.e. Agomelatine. The design system increased the bioavailability by 2.82 folds. [21] Oral bioavailability of different types of drugs improved up to 10-20 folds by using the hydrophobic ion-pairs (HIPs) incorporated into the lipid-based nanocarrier systems thereby enabling the active molecules to cross the phospholipid bilayer of epithelial cells for an effective therapeutic intervention. The intestinal drug solubility is an important parameter in the drug profile. Some poorly water soluble and lipophilic drugs have the disadvantage of low bioavailability. These issues of aqueous solubility and gastrointestinal absorption can be effectively enhanced by making use of the lipids. Due to the incorporation of the drug into the lipid-based nanocarrier systems the stability of the drug into the gastro-intestinal (GI) track is substantially enhanced, [22] while the bioavailability of the hydrophobic drugs have also been improved by using solid lipid particles (SLPs). [23] The literature bears evidence that there is an increase in the oral bioavailability after the use of solutions, suspension and self-emulsifying formulations based on the lipids as well as LBDDS. [24]

4.2   Encapsulation

Drug encapsulation is the process in which the active pharmaceutical ingredient is entrapped by the carrier material i.e. lipids to form suitable particles. [25] Successful encapsulation process of active compounds depends upon the selection of the suitable encapsulating agent. Various types of these materials are used for the encapsulation process. In an aspect of functionality, encapsulating agent must possess following properties such as it should be a good emulsifier, control over viscosity in high concentration, adequate dissolution and network-forming properties, ability to preserve and protect active compounds at different conditions, capability to infiltrate adverse condition such as acidic and enzymatic barrier, more precise in target selection, ability to increase the adherence capability or residence time of active compounds in target sites.

Other than lipid, polysaccharides and proteins can also be used as an encapsulating agent. Since the only focus of this chapter is lipid based drug delivery systems, this section is limited to lipid based carrier agents only. The attainment of any biocompatible lipid based system is dependent on the encapsulation efficacy as well as the drug release profile. [26] Lipids-based carriers have outstanding performance in emulsification, film formation as well as encapsulation of active compounds. Fats and oils, which consist of both polar (e.g. monoglycerides, phospholipids) and nonpolar (e.g. triacylglycerol, cholesterol) are the best representatives that fall in this category of encapsulating agents. [25,27,28] Due to the biphasic nature of lipid, LBDDS has the ability to encapsulate both lipophilic and hydrophilic drugs. [25,28] Water soluble drugs occupy the aqueous core region whereas lipophilic drugs are generally encapsulated in the lipid bilayers of LBDDS. Efficiency in drug loading capacity of the LBDDS depends upon the various factors such as use of appropriate cross-linking agents (glutaraldehyde, formaldehyde, carbodiimide) or physical treatments (i.e., UV irradiation, freeze-drying) during their preparation process. [26] Generally, LBDDS are prepared by the hydration of a dry lipid film or precipitation of lipids or adsorption of dissolved lipids at liquid interfaces. The hydrophilicity or hydrophobicity index of the drugs will decide the solubility of drugs either in in the aqueous medium (hydrophilic) or in the lipids (hydrophobic). [29] Thus encapsulation improves bioavailability, controlled release, and targeting precision of active compounds. [25]

Functional foods are specially designed and introduced in day to day life to improve human health, well-being, and performance. This functional food contains the phenolic compounds, vitamins as well as minerals, but suffers from poor solubility, unwanted flavor/smell, instability and low bioavailability profiles. The above said problem of the functional food can be overcome by the encapsulating bioactive components into the suitable system such as nanoparticle-based delivery systems or LBDDS by using the lipids. [30,31] According to the World Health Organization, about 80% of the population in the developing country depends on medicinal plants for their primary health care. Various types of plants based active moieties have been produced through diverse approaches. The bioactive herbals have several health benefits but less bioavailability, short half-life; low membrane permeability due to the large molecular size are the key concerns for larger public utility. Through their encapsulation within lipid systems, important parameters like solubility, absorption, target specificity, cell permeability can be improved significantly. [32]

4.3 Toxicity

Toxicity of the active pharmaceutical ingredient (API) can be minimized by using LBDDS. Use of advance type of the lipid based system, known as bilosomes, aids in the improvement of the bioavailability as well as reduces the toxicity associated with the API. Elnaggar and group were the first to work on the enhancemnt of the oral bioavailabilty and minimise the oral toxicity of the active pharmaceutical ingredient i.e. Risedronate (RS). It was reported that when RS encapsulated with bilosomes, its bioavailability increased by 1.5 times whereas toxicity is reduced by 2 folds. [33]

4.4 Biodegradability

Use of biodegradable materials for biomedical applications has witnessed enormous upsurge because it enhances the biocompatibility and also facilitate elimination once their role is over. [34] In order to improve the biocompatibility, bioavailability, safety, permeability, minimize toxicity and achieve better retention time, the use of lipids and polymers is on the rise as they can vest biodegradable properties to the therapeutic payload and the LBDDS as a whole. In this regard, it becomes essential to design appropriate nanoparticle formulations which are biodegradable in nature, for the safe and efficient transport as well as release of the drug selectively to the target site. Selection of suitable biodegradable polymer/lipid as the drug-carrier for the formulation of LBDDS can also be based on the target organ compatibility, location and physiology.[35]

5. Lipid Composites

Conventional LBDDS are generally delivered orally, while they passively accumulate at the diseased areas through enhanced permeation and retention (EPR) effect. Although the conventional lipid possesses many prominent features required for drug delivery system but still it is far from ideal DDS in terms of bioavailability and target specificity and hence there is lot of room to improve present LBDDS for accomplishing a better therapeutic effect. Properties such as drug solubility, dispersion, digestion, absorption, morphology, irritancy, toxicity, controlled release and target specificity are needed to be addressed using new methods or technology. To meet these requirements of modern DDS, LBDDS need to be modified with different materials possessing different physical, chemical or biological properties (Figure 3). Formulation of LBDDS with one or more of these constituents having sizes in 'nm range' generates a composite with new sets of additional properties, known as lipid composites (Figure 4). This nano-composite may offer many advantageous over conventional LBDDS such as improved bioavailability and biocompatibility, target specificity, drug release profile as mentioned above. [26,36] Based on the applications, the lipid composites are broadly classified into stealth liposomes, targeted liposomes, stimuli sensitive liposomes, and some miscellaneous and specific liposomes such as neosomes, transferesomes, ethosomes and liposomal emulsion (Table 1).

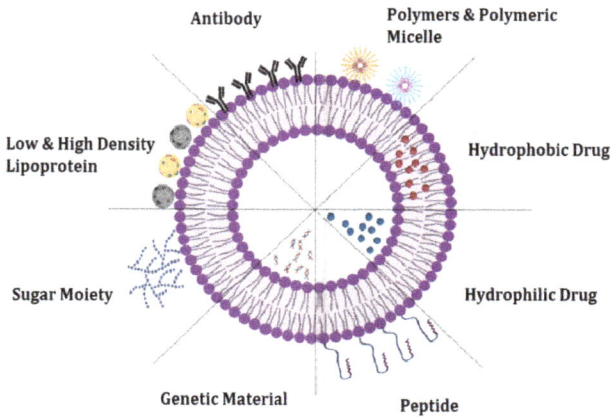

Figure 3: Different liposomal regions for targeted delivery.

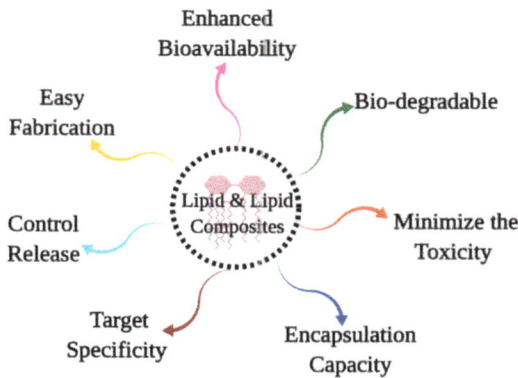

Figure 4: Different features of lipid and its composites for drug delivery system.

Table 1: Lipid composites used in various Drug Delivery Systems

Type of Drug Delivery System	Drug + Polymer/lipid + Lipid Composites/System	Method of Preparation	Developed for	References
pH sensitive liposomes	Doxorubicin + PEGylated phospholipid + tumor-triggered tabletargeting ammonium bicarbonate liposomes	Film hydration Method	Tumor-Specific Drug Delivery	[37]
	Catalase + DOTAP+ Cationic lipid-coated magnesium phosphate nanoparticles	Micro-emulsion precipitation.	Intracellular Protein Delivery	[38]
	Doxorubicin+ DOTAP+ loaded cationic lipid	Thin film hydration and ultrasound method	Suitable as potential low toxicity cationic drug delivery systems	[39]
	Gemcitabine + DOPE + Hyaluronic acid functionalised pH-Sensitive Liposomes	Thin film hydration method	Overcome Gemcitabine Resistance in Pancreatic Cancer	[40]
	Sorafenib + VEGF-siRNA+ DOTAP + Carboxymethyl chitosan-modified liposomes	Thin-film hydration method	Synergistic treatment of hepatocellular carcinoma	[41]
	Paclitaxel + PEG-PLL-DMA + Charge-reversal and NO generation cationic liposomal system	Thin-film hydration method	Multidrug resistance in cancer	[42]
Thermo-sensitive Liposomes	Ovalbumin + DOTAP + DPPC+ liposome-in-gel carrier by heterotopic mucosal engrafting	Film hydration method	CNS delivery of proteins	[43]

	Doxorubicin + DPPC + DSPE-PEG$_{2000}$ + Lysolipid-based thermosensitive liposomes	Lipid film hydration and extrusion method	Tumor-target against cancer therapy	[44]
	Doxorubicin + PLGA-PEG-PLGA based thermogel	Film dispersion technique	Sustained local drug delivery for the treatment of breast cancer	[45]
	Doxorubicin + DPPC+ DSPE-PEG$_{2000}$ + Quick-Responsive Polymer-Based Thermo-Sensitive Liposomes	Film hydration method and solvent evaporation method	Controlled Doxorubicin Release for cancer therapy	[46]
Magnetic Liposomes	Irinotecan + DPPC + Thermosensitive Magnetic Liposomes for Alternating Magnetic Field	Thin-film hydration method	Dual Targeted Brain Tumor Chemotherapy	[47]
	Geldanamycin + DPPC + DSPE-PEG$_{2000+}$ Folic Acid-Targeted Temperature-Sensitive Magnetoliposomes	Rotary evaporation method	Ovarian cancer	[48]
	Curcumin + polyethylene glycol + Magnetic nanoparticles	Co-precipitation method	Hydrophobic anticancer drugs	[49]
Solid Lipid Nanoparticles	Nucleic acid + cholesteryl oleate + cationic SLN	Hot microemulsification method	Gene therapy applications	[50]
	Ofloxacin+ chitosan and polyethylene glycol (PEG) coated SLN	Modified emulsion/solvent evaporation method	Ocular Delivery	[51]
	Etoposide + Tween 80 + SLN	Melt-emulsification and ultrasonication technique	Ocular Delivery	[52]
	Ofloxacin + Pegylation and Chitosan Coated +	Modified emulsion/solve	Ocular Delivery	[51]

	SLN	nt evaporation method		
Transfersomes	Natamycin natamycin + Gellan gum+ sol-to-gel transforming transfersomes	Film hydration technique	Topical ocular delivery	[53]
	Colchicine + β-cyclodextrin + transfersomes	Lyophilization technique	Transdermal delivery	[54]
	Human growth hormone + Lecithin soybean phospholipid + nano-transfersomes	Modified thin-film hydration method	Transdermal delivery	[55]
Niosomes	Plumbagin+ hydroxyl propyl betacyclodextrin + niosomes	Lipid layer hydration method	Antifertility activity	[56]
	Plumbagin + beta cyclodextrin + cholesterol, Span 60 and dicetyl phosphate + niosomes	Lipid layer hydration method	Anti-tumour activity	[56]
Emulsomes	Oxcarbazepine + PLGA-PEG-PLGA Emulsomes Composite	---	Brain Delivery via the Nasal Route	[57]
	Oxcarbazepine + soya phosphatidylcholine + Nano-spherical emulsomes	Modified thin film method	Brain Delivery via the Nasal Route	[58]
	Sparfloxacin + Phospholipon 90G + thermosensitive emulsomes	Thin film hydration technique	Ophthalmic delivery	[59]
Drug conjugates	Gantrez® AN-thiamine polymer conjugate+ zein nanoparticles	Desolvation technique	mucus-permeating properties	[60]
	CXCR4 antibody drug conjugates	---	Anti-tumour activity	[61]
	Hypericin-hydroxypropyl-b-cyclodextrin inclusion complex + liposome conjugates	Thin film hydration method	Antitumor and Antiangiogenic activity	[62]

	Paclitaxel + siRNA + polyethyleneimine-modified liposomes	Thin-film formation method	Drug-resistant cancers	[63]
Stealth liposomes	Protein corona + DSPC/cholesterol/DSPE-PEG2k + antibody conjugated Stealth liposomes	Extrusion method	Antibody targeting	[64]
	Calcein + Pegylated (stealth) liposomes conjugated to human serum albumin	Modified lipid film hydration method described	Breast cancer therapy	[65]
	Docetaxel + RIPL peptide-conjugated liposomes	Thin-film hydration method	Targeted drug delivery in blood circulation	[66]

DOTAP - 1,2-dioleoyl-3-trimethylammonium-propane
DPPC - Phospholipids 1,2-dipalmityol-sn-glycero-3-phosphocholine
DOPE - 1,2-Dioleoyl-sn-glycero-3-phosphoethanolamine
PEG-PLL-DMA - Monomethoxypoly (ethylene glycol)-poly(L-lysine)- graft-dimethylmaleic anhydride
DSPE-PEG$_{2000}$ - 1,2-distearoyl-sn- glycero-3-phosphoethanolamine-N-PEG$_{2000}$

PLGA-PEG-PLGA - poly(lactic acid-co-glycolic acid)-poly(ethylene glycol)-poly(lactic acid-co-glycolic acid)

5.1 Stealth Liposomes

The conventional or first generation liposome, typically composed of lipid and cholesterol suffered from: rapid clearance from the blood stream, inability to evade the immune system, steric stability and healthy cell cytotoxicity, in case of charged lipid. The distribution of drug in other site of the body is impeded due to rapid absorption by the mononuclear phagocyte (MPS). The conventional liposomes can come to rescue only when it is the case of delivery of anti-parasitic and antimicrobial drugs, because the target is localized in MPS. However, if the target is beyond the MPS then this behavior becomes the drawback of conventional liposome for drug delivery. To overcome these problems, a new approach is introduced wherein surface of liposome is modified with an inert hydrophilic polymer. Such surface modified liposomes falls in the category of 'stealth liposome'. [67,68] There are several materials that can be used to modify the surface of liposome such as glycolipids or sialic acid, but the most famous and successful formulation is with PEG of varying chain lengths (Figure 5). The PEG coated liposomes have shown prolonged blood-circulation time and low absorption by MPS. Besides this PEG possesses many useful properties, such as biocompatibility, solubility in aqueous and organic media, very low immunogenicity, low toxicity, easy excretion from the body, and moreover it is easier to conjugate with the lipid. The presence of PEG on liposome surface also helps to control the aggregation of the vesicle and increase the stability by reducing van der Waals forces among the vesicles. To increase the specificity of stealth liposome, terminal PEG can be further modified with monoclonal antibodies or ligand. One of the well-known formulation utilizing stealth liposome is of PEG grafted liposomal doxorubicin which is also approved by both the Food and Drug Administration (FDA) and European Federation. Despite the numerous important topographies of stealth liposomes, insufficient unloading of drug and more importantly lack of target specificity are the major concerns. Stimuli responsive or target based liposomes, which possess certain properties that will only allow release of drug at specific site, can be the alternate drug delivery vehicle to overcome the issues faced by conventional or stealth liposome.

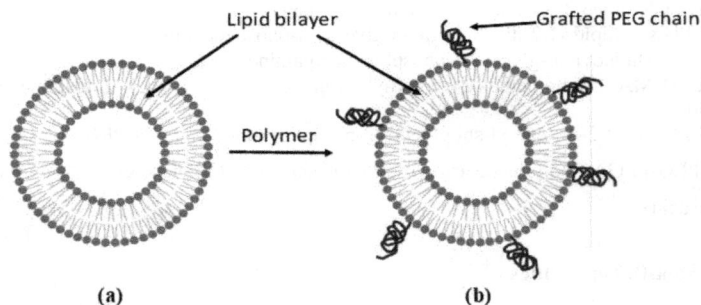

(a)   **(b)**

Figure 5: Schematic representation of conventional (a) and stealth liposome (b).

5.2   Stimuli Sensitive Liposome

Although conventional liposome and stealth liposome shows much encouraging result in terms of reduced toxicity, long circulation, reduction of macrophage uptake, but still EPR effect of drug is a major problem since liposomes not only accumulate in abnormal cell but also in normal cells which may turn a lifesaving drug into of life taking one. To increase the specificity of drug delivery systems, a new line of attack (referred to as active attack) is used, known as stimuli-sensitive liposomes as pharmaceutical carriers. A stimuli-sensitive liposome is a category of liposome that mostly depends on external or internal factors in order to trigger the release of the drug, protein and gene. [69,70] The stimulating agents in this case, can be the pH, light, magnetism, temperature, specific enzymes or ultrasonic waves. This method is now gaining popularity among other smart delivery systems particularly in antitumor therapy where side effect of chemotherapeutic agent is a major concern due its non-specificity towards the targeted tumor cells. The tumor microenvironment being characterized by acidic pH, hypoxia, or unique enzymatic activity provide more favorable conditions to develop stimuli sensitive liposomes. Lipid can be impregnated with stimuli-sensitive component to realize triggered sensitive liposome. Stimuli sensitive release is usually based on the destabilization of lipid bilayer as consequence of stimuli triggered reaction. This section will focus on liposomal drug delivery approaches using pH change, heat, light, magnetism, ultrasound etc. as external and internal stimulus.

pH Sensitive Liposomes: pH sensitive liposomes are composed of phospholipid bilayer that are sensitive to the change in pH. Since some pathological sites/cell possesses different pH than a normal cell, hence it can be used as stimuli to trigger the drug unloading from the liposome. The pH alterations in abnormal tissues causes protonation/deprotonation of lipid head groups which affect permeability of the liposomal membrane as a consequence of morphological changes in the lipid bilayers. This change in the lipid bilayer due to influences of the pH releases the drug from liposomal encapsulation. pH sensitivity is the important feature of this type of the liposome which gives advantage for the delivery of the drug content at various possible targets. The idea of the development of the pH sensitive drug delivery is coined in 1980 by the Matsuzaki, Hamasaki and Said. [71] The main objective of this drug delivery system is the unloading of liposomal contents in response to a low pH environment especially in case of cancer therapy as well as gene delivery. This system has the capacity to interact as well as enhance the fusion or disruption of the endosomal membrane and perfectly release the encapsulated drug in to the cell cytosol. [72] The development of pH sensitive liposomes is attained by two main strategies: 1) By making use of materials which can undergo conformational and/or solubility changes in response to change in local pH; 2) pH labile bond which releases the attached molecule or drug from the polymer backbone. [73,74]

The most common lipid carriers used for the preparation of the pH sensitive liposomes is 1,2-diole-oyl-sn-glycero-3-phoshoethanolamine (DOPE) and 1,2-dipalmitoyl-sn-glycero-3-phospho-ethanolamine, which under acidic conditions are transformed from a lamellar phase (L_α phase) into a fusogenic hexagonal phase (H_{II} phase). The pathological inflammated, infectious and cancerous tissue has the different pH profile as compared to the normal tissues. The pH of tumor cell is often more acidic due to accumulation of lactic acid in hypoxia condition (Warburg effect), than normal cell and sometime it may go down to 5.6. Therefore, taking the advantage of the tumor microenvironment, novel drug delivery systems can be developed to deliver the drug at the diverse areas where variation in the pH is regularly observed. Therefor the anticancer drugs and gene delivery are principle areas of interest for the development of the such type of novel drug delivery systems.[75] The applications of the pH sensitive systems are multipurpose which includes tumour targeted drug delivery, tumour diagnosis, chemotherapy, vaccine delivery as well as gene delivery. [72] However, pH triggered delivery suffered from two major limitations, one because of remotely placed acidic sections of tumors, makes it difficult efficiently accumulate in targeted site. Secondly, tumor pH rarely reaches below 6.0 that makes it difficult of design liposomes which work in small pH range ($\sim$ 1).

Thermo-sensitive liposome: All the lipids are characterized by a specific temperature (T_m) where they change their phase from solid (gel) into liquid crystalline phase. At this temperature the hydrocarbon chains are haphazardly oriented and behave like fluid which results in rapid change in the location of lipid molecule with their adjacent/ immediate neighbors. This behavior of liposome is being used in drug delivery for the last 40 years in the name of thermos-sensitive liposomal drug delivery. Thermosensitive liposomes (TSLs) are drug delivery systems for targeted delivery that discharge the encapsulated drug at elevated temperatures ($\sim$40–42°C). Thermosensitive lipids were globally investigated since 1978 and various studies have been documented for the same. These liposomes are the most reliable systems for the treatment of the various diseases as well as diagnosis because of their capacity of site specification due to the response at certain temperature which is the main advantage of the site-specific delivery. [76] The site specificity reduces the exposure of the drug to the healthy tissue which leads to amplification in the therapeutic value. Drug moiety encapsulated into the thermosensitive liposomes are stable at the physiological temperature because there is no unloading of the drug in the normal/healthy tissue. [73,77] Liposomes have been used for cancer treatment since 1990s, with primary objective to minimize the toxicity of anticancer drugs as well as to enhance the bioavailability. Thermo-sensitive liposomes are especially formulated to fulfill the above said criteria. They release the encapsulated active moiety after the expose to the hyperthermia temperatures i.e. 40–42°C. Due to localized hyperthermia which is assisted by an image-guided hyperthermia device, TSLs allows the site specification/targeted drug delivery. Recent formulations of TSLs are permitting the drug uptake up to 20-30 times higher as compared to the free drug (i.e. without encapsulation). [78,79] Figure 6 shows the structures of some lipids used in the preparation of thermo-sensitive liposomes.

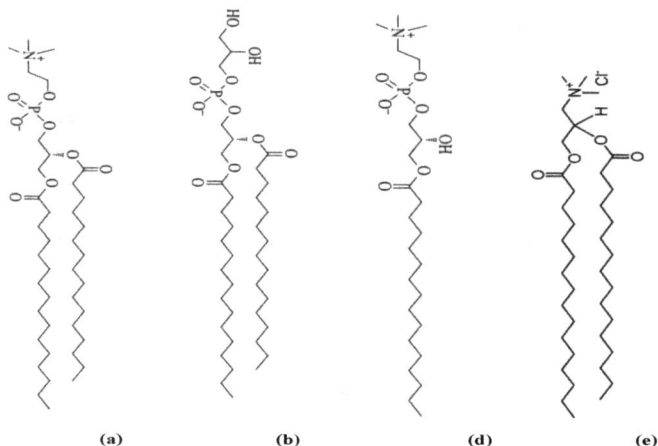

Figure 6: Chemical structure of some TSL : a) 1,2-dipalmitoyl-sn-glycero-3-phosphocholine, (b) 1,2-dihexadecanoyl-sn-glycero-phospho-(1'-rac-glecero) sodium salt, (c) 1-myristoyl-2-palmitoyl-sn-glycero-3-phosphocholine, and (d) 1, 2 dipalmitoyl-3-trimethylammonium-propane.

The prime concern is the stability of the formulation being much less therefore there is a chance of leakage of the drug at 37 °C in the bloodstream which may increase the risk of toxicity. [44] The stability of the temperature sensitive lipid could be increased or decreased by controlling many factors such as hydrocarbon chain length, unsaturation, charge, and the type of headgroup species. Increase in the hydrocarbon chain length may increase the stability of liposome due to enhanced van der Waals interaction. Introduction of unsaturation produce a kink in the hydrocarbon chain which reduces the compactness of the liposome and subsequently the transition temperature. [80] To overcome this problem, optimized polymer-based thermo-sensitive liposomes (P-TSLs) can also be used for the site specific controlled release of the drug like doxorubicin. [46] Cao and co-workers prepared the hybrid system of liposomal doxorubicin loaded thermo-gel (DOX-Lip-Gel) for the sustained discharge of the doxorubicin for the treatment of the breast cancer locally. The prepared formulation being liquid at room temperature is switched into a gel state at body temperature. The results describe that, there is a better release profile and lower side effects by the released doxorubicin when compared to the conventional system. [45]

Other-stimuli sensitive liposome: In addition to the pH and heat, there are other stimuli which can be used as triggering point for the drug release from liposomal encapsulation such as light, magnetism and ultrasonic waves. Light can also be used as a physical impetus to trigger the release of liposome-encapsulated molecules at the targeted sites.[81–84] Light triggered liposome can be prepared by introducing additional photo-reactive group or molecule in head group, glycerol part or in fatty acyl chain of lipid. In addition to the photo-reactive group, the wavelength, intensity and exposure of light also play asignificant role in light directed DDS. Generally light in the range of 600–900 nm wavelength is known as therapeutic window since this range is less hazardous to the tissue and it is transmissible deep into biological tissues.[85] Light triggered destabilization of liposome may involve either degradation of light sensitive lipids,[86,87]photosensitization by water soluble probe (photosensitizer, PS), photo-polymerization[88,89] or by photo-oxidation.[90] Incorporation of photo-labile lipid molecules in conventional liposomes, destabilizes the lipid bilayer in presence of suitable light (Figure 7). For this purpose, photo labile group such as nitropyridine, dihydroxybenzophenone based lipid can be used in conventional liposome formulations for photo-triggering.[91] Most of the photo-cleavable lipids reported in the literature absorbs only short wavelength light to get activated, which limit their application as actual DDS.

The liposomal content can also be released by incorporating green/red light absorbing dye (PS) into lipid bilayer, which destabilize lipid layer either via direct energy transfer from PS to lipid or generation of reactive oxygen species (ROS). PS based light sensitive liposome can also be used in photodynamic therapy (PDT), a light based cancer treatment, to increase the stability and tumor selectivity. The PDT PS such as Chlorin e6, meta-tetra[hydroxyphenyl]chlorin, phthalocynines and metallophthalocyaninesare used in liposomal formulations to enhance the cytotoxicity, target specificity and to reduce the side effect. Even though the number of photosensitive liposomes are formulated but theirpractical application remains limited due toincapability of visible radiation to penetrate into biological tissues. To some extent this problem can be overcome with help of nanocomposites made up of photo-responsive liposome built with gold nanoparticle. This new nanocomposites uses near infrared light as light source which can undergo deep tissue penetration with minimal phototoxicity.[92]

Advanced Applications of Bio-degradable Green Composites   Materials Research Forum LLC
Materials Research Foundations **68** (2020) 156-195   https://doi.org/10.21741/9781644900659-7

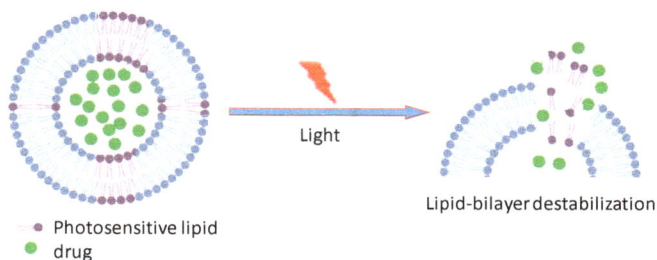

Light

Lipid-bilayer destabilization

● Photosensitive lipid
● drug

Figure 7: Graphical illustration of light-triggered drug delivery strategy.

Another example of stimuli liposomal nanocomposite is magnetic nanoparticles embedded in liposomes know as magnetic liposomes or magneto-liposomes. Magnetic nanoparticle in drug delivery system not only guided the liposomes towards its target in external magnetic field but also used for MRI imaging and heat generation. In presence of high frequency external magnetic field, liposome containing magnetic nanoparticle could be directed to the targeted site. It has been reported that hyperthermia induced by local magnetic field are able to reduce cell viability drastically.[93]

From last two decades, application of high intensity, low frequency ultrasound waves have been used for controlled drug release from liposome. Ultrasound has some advantage over light as liposomal stimuli because of non-invasive nature, high permeability through blood–tissue barriers and cell membranes as well as deep penetration into interior of body, therefore it can be used to transmit energy into the body at precise locations. It has been found that liposome containing air bubblescan potentially respond to the ultrasound stress (created by ultrasound waves) by releasing encapsulated payloads. The bubbles can be produced while preparing liposome using ultrasonication method commonly known as cavitation which also causes their oscillation. If the shear stress formed by oscillating microbubble, due to ultrasound waves, is more than the strength of the vesicle, it will break and discharge it contents. The focused ultrasound waves can create significant amount of thermal energy which causes solid order (SO) to liquid order (LO) phase transition of liposome. The SO to LO phase transition disrupt the packing of the lipid bilayer which permit the drug to move across the lipid bilayer.

5.3   Solid Lipid Nanomaterial

Although, the liposome is the pioneer of nanoparticle based drug delivery systems but the control over the drug release and delivery are the big issues yet to be tackled. Solid lipid nanoparticles (SLN) or lipospheres has been explored by the researches around the globe for their potential/versatile applications in the principal area of the pharmaceutical sciences.[94–96] SLNs **(Figure 8)** have been studied for controlled release of drug, and to increase bioavailability of certain drugs.[97]SLN have significant advantages over the other colloidal systems, such as good permissibility, biocompatibility and ease of scale-up. Size dependent properties of SLN makes it more appealing for various therapeutic applications. SLNs is formulated of solid lipids such as di- (glycerol bahenate) or triglycerides (tristearin), glyceride mixtures or waxes which is stabilized by biocompatible surfactants/emulsifiers(Figure 9**)**. Surfactant used in SLN can be further grafted with different functionalities. SLNs could be an alternate drug delivery system to that of delivery strategies discussed in the above section. It offers numerous advantages such as targeted drug release, high drug loading capacity, competent in carrying both hydrophilic and lipophilic drug, water based formulation, lower the prevalence of acute or chronic toxicity, improved drug stability, excellent biocompatibility, less expensive, easy to scale up and sterilize.

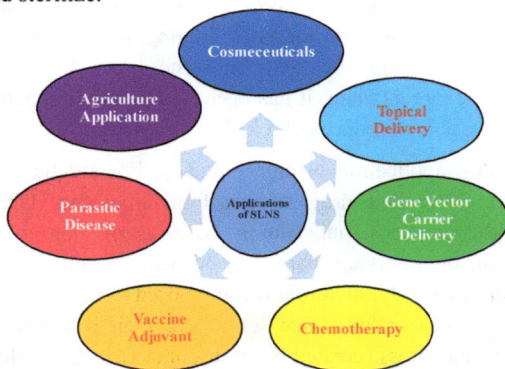

Figure 8: Diverging radial diagram showing various applications of SLN system.

The preparation of SNL involves very high pressure and rapid temperature variations that also increase the risk of drug degradation, gelation phenomena lipid crystallization, and the formation of other colloidal species. Due to temperature changes during the formulation process, chances of a highly ordered crystal lattice formation of lipid matrix is more which may reduce the control on drug release rate. The choice of lipid which does not form good crystals such as complex lipids (mono- or di-glycerides, or triglycerides) with different chains lengths, may minimize this problem.

Nanostructured lipid carriers (NLC) epitomize a novel and developed generation of SLNs which are formulated by using solid lipid matrix entrapping liquid lipid nano-compartments which become solid at body temperature.[98]This new generation lipid carriers (NLCs) is innovated to minimize the problems accompanying with SLN such as limited drug loading capacity, drug eviction during storage and low consistency in drug release rate, low shelf life, etc.[11,98–100] Three models of NLCs have been proposed so far: imperfect type, multiple type and amorphous type NLC, on the basis of solubility of the drug, lipid content, encapsulation capacity and the particle size of the final product. Formulation procedures for both SLNs and NLCs are identical. The melted solid or semi-solid lipid blend is mixed with API in the presence of surfactant/stabilizer solution at lipid melting temperature and high pressure. The oil-in-water (o/w) nanoemulsion obtained after homogenization of above mixture, forms crystalized emulsion droplets after cooling. Depending on the starting material, either SLN or NLC, the crystallized emulsion droplets forms lipid nanoparticles with solid particle matrices.[11,98–100]

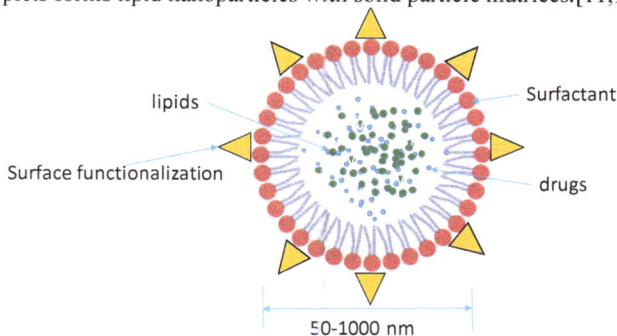

Figure 9: Schematics of Solid Lipid Nanoparticles.

Nutlin3a has an inhibitory effect on the function of P53 protein (regulates apoptosis mechanism in the cell) and also enhances the therapeutic effect of chemotherapeutics, but the oral bioavailability as well as stability of the Nutlin3a is very less. This specific problem is overcome by using SLNs (Nutlin3a-containing cationic solid lipid nanoparticle) formulation which appears to be the best possible option for the degradation of the P53 protein.[101]Liu and Zhao prapred all types of the SLNs i.e. Polymer lipid hybrid nanoparticles (PLA NPs), solid lipid nanoparticles (SLNs) and nanostructured lipid carriers (NLCs) for the Tetracaine (TTC) loaded nano-anestheticsystems. Comparative studies of the all nanoparticles indicates that, TTC-NLCs composites are the promising system for the long-term anesthesia.[102] To reduce the toxicity of gene therapy, cholesteryl oleate as a novel excipient for SLNs formulation loaded with nucleic acid are used.[50]Nano-scale drug carriers is the best method for delivery of active ingredient at cellular level for cancer therapy. Celecoxib (CLX) loaded Solid lipid nanoparticles (CLX-SLN), nanostructured lipid carriers (CLX-NLC) and a nano-emulsion (CLX-NE) designed and formulated by Üner, Yener and Ergüven for the treatment of breast cancer and acute promyelocytic leukemia. Nanoformulation of the CLX achieved higher CLX efficiency at cellular level as compared to the conventional system. The results suggest, prepared formulation is administered along with combination therapies for the treatment of above said diseases.[103]The oral bioavailability of the asenapine maleate (AM) (Antipsychotic agent) has been amplified by 50 fold after fabrication in SLN system which open up new avenues for the treatment of schizophrenia.[104] For the treatment of the oral squamous cell carcinoma (OSCC), combinatorial approach has been implemented by Bharadwaj and cowokers where paclitaxel (PTX), 5-fluorouracil (5-FU) and ascorbic acid (AA) used as the active ingredients for fabrication of the SLNs. The pharmacokinetic and biodistribution studies of prepared SLN shows better therapeutic efficacy.[105]

Distribution coefficient of hydrophilic drugs severely affects the loading capacity of SLNs. Hence only highly potent low dose hydrophilic molecules are suitable for SLN composites. This issue can be addressed by using lipid-drug conjugate nanoparticle (LDC) in which drug molecules are chemically modified with lipid.[10,106,107]In addition to this LDC improved targeting to the lymphatic system and showed enhanced tumor targeting. Based on the problems in drug delivery, site of delivery and chemical constituents of drugs and lipids, various conjugation strategies can be used to synthesize LDCs. It is class of prodrug where pharmaceutically active molecule is linked with lipidvia covalent or non-covalent bond, which is further processed under high pressure homogenization to nanoparticle in presence of aqueous surfactant. Drug-lipid conjugate can be linked via various types of bonds for instance ester, amide, hydrazone or disulfide bond (Figure 10).All these bonds are labile to bond scission certain conditions. Hence whenever these lipid-conjugates are introduced in such circumstances in patient body, it gets metabolized and releases the active moiety. Various lipid derivative can be used as conjugate such as fatty acid, phospholipids, glyceride steroids etc. The choice of lipids is simple based on their biocompatibility, necessary functional group to conjugate drug, safety, self-assembly nature, transient temperature (T_m) and biodegradability. Ester linkage which is formed by hydroxyl group of drug and carboxylic group of lipids, can be specifically broken by enzyme such esterase whereas disulfide bond can be cleaved under reducing environment to release drugs. The amide bond between drug and lipid can be formed by 'carbodiimide coupling' reaction. Lipophilic drug remains inactive until amide bond is cleaved to free the active drug. Comparatively ester bond, amide linkage is much stable that result in slower drug release rate. Hydrazones based lipid-drug composites are pH sensitive. It remains stable at neutral pH, whereas undergoes degradation at a lower pH. Such type of lipid-drug conjugate has been successfully utilized for the delivery of chemotherapeutic drugs. Due to increased lipophilicity of conjugated drug, poor absorption and premature hydrolysis can be avoided. Lipid conjugates are particularly popular for delivering chemotherapeutic agent due its target specific activation which reduce the side effect of chemotherapy. Crossing blood brain barrier (BBB) is again a big challenge for many active drugs. BBB only allows uncharged small molecules (MW < 500) having within range o/w partition coefficient (<5). This barrier limits the usage of large but potent molecules having low lipophilicity. LDC can enhance the brain delivery of larger molecules by targeting different lipid receptors in BBB.[10]

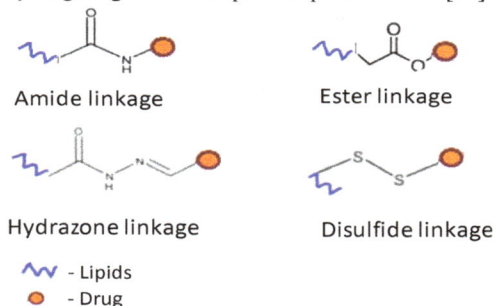

Figure 10: Different chemical bonds used as linkage in the construction of lipid-drug conjugates.

6. Other Types of Lipid Composites

Transfersomes: The drug delivery platforms developed so far were inefficient to overcome the limitations associated with orally administered drugs as well as in parenteral preparation and so scientists searched for the other routes of administration with the high therapeutic activity as well as patient compliance. Skin is the largest organ of the body having surface area in between 1.5 to 2 m^2 in an adult human and is almost directly accessible to many organs that make it an attractive target for the advancement in topical drug delivery. In addition to this other important applications of skin administration are: avoidance of adverse systemic effects of drug, avoid the gastrointestinal degradation and more importantly easy to introduce and high patient compliance. Topical drug delivery has also some disadvantages like local irritation effect, erythema, itching, and low permeability in the stratum corneum, and it is only useful for spot specific relief etc. In this regard, a new type of vesicular drug carrier system called transfersomes [108–110] shows potential to overcome aforementioned obstacles. The term transfersomes and concept coined in 1991 by Gregor Cevc [111]. Transfersomes are stress-responsive, elastic and extremely adaptable aggregate contains an aqueous core enclosed by lipid bilayer. German Company IDEA AG, registered the 'Transfersome' as a trademark for patented drug delivery technology. The term Transfersome derived from the Latin word, "transferred" meaning "to carry across," as well as Greek word "soma," for a "body" i.e. carrying body (artificial vesicle). Although it looks same as that of conventional liposome, but transfersomes are highly flexible and can deform and reform wherever it passes through pore which is much smaller (sometimes 10th time smaller) than its size (Figure 11).

Transfersome are deemed with many advantages such as, patient compliance, sustained release of the drug for the longer period of time which leads to reduction in dosing frequency, equivalent therapeutic effect, utilized for such type of the drugs which have the narrow therapeutic window, improved bioavailability and flexibility for the drug administration. Transfersome is composed of two main components in different ratio, amphipathic ingredient (phosphatidylcholine) which is responsible for the self-assembly and a bilayer softening component (such as a biocompatible surfactant viz sodium cholate, sodium deoxycholate, Tween 20, 60 and 80, Span 60 65 and span 80) which is responsible for flexibility of the transfersome.[112] Transfersome can be prepared by vortexing-sonication method, suspension homogenization process, modified handshaking process, aqueous lipid suspension process, centrifugation process, thin film hydration technique, etc. [113].Transfersomes can utilized for the delivery of various agents such as insulin, corticosteroids, proteins and peptides, interferon (INF), anticancer drugs, anesthetics, non-steroidal anti-inflammatory drugs (NSAIDs) as well as herbal drugs.[111,114] Waheed and co-workers optimized the raloxifene hydrochloride-entrapped, limonene containing transfersomes. The optimized system showed high bioavailability as compared to the oral suspension of the raloxifene.[115] For the treatment of rheumatoid arthritis (RA), long-term treatment can improve patients' compliance and reduce the accumulation of drug side effects. For the same, Song and research group prepared sinomenine hydrochloride-loaded transfersome which enhance transdermal permeability and drug deposition for the oxidant stress of RA.[116] Alternative for the local anesthetic injection were discovered by the Omar *et al.* by formulating the topical gel containing transferosomal lidocaine. Transfersomes containing preparation enhanced skin permeation as well as increased the local anesthetic efficacy as compared to control formulation.[117]

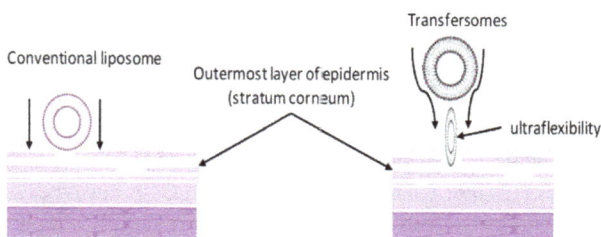

Figure 11: Schematic diagram show penetration of conventional liposome and transfersome.

The low bioavailability colchicine due to ionization at gastrointestinal pH may lead to some adverse side effect and can also be overcome by using transferosomal formulations.[54]Along with advantages, transfersomes has few disadvantages like, physiological barrier of the skin which changes from person to person according to age, potency of the drug molecule, applicability may vary with patch type and environmental conditions, chances of the skin irritation and hypersensitivity reactions, less stability due to oxidative degradation along with high cost of the product.[118]

Niosomes: Niosomes are multilameller vesicular structure of nonionic surfactants, such as polyoxyethylene alkyl ethers or esters, and excipients such as cholesterol resulting in closed bilayered structures.[119] Niosomes were first time investigated in the 1970s as a better tool for the cosmetic industry but due to many attractive features, they have been also explored as drug delivery vehicles in 1980s.[120] Primarily because of the use of the more stable and significantly affordable non-ionic surfactant formulations, niosomes are widely used and investigated as an alternative to naturally occurring liposomes.[119,121] The advantages of niosomes as drug delivery vehicle[119] include, sustained drug release profile, targeted drug delivery, high therapeutic efficiency with minimum drug dosage, amphiphilic in nature which leads to delivery of both hydrophilic and lipophilic drugs, better patient compliance compared to oily dosage forms, improve the oral bioavailability of poorly soluble drugs as well as enhance the skin permeability of drugs when applied topically, good encapsulation capacity, biodegradable, biocompatible and non-immunogenic in nature.

Niosomes are mainly composed of surfactant and additives. Various types of the surfactant molecules are reported[122–124]for the preparation of niosomes which have the capacity to entrap the hydrophilic and lipophilic material Non-ionic surfactants used in the niosomes are amphipathic, including terpenoids (Squalene), spans, polysorbates, alkyl oxyethylenes (usually from C_{12} to C_{18}), etc. The additives used in niosome preparation are generally cholesterol and the charged molecules, where the later gives rigidity to the bilayer as well as improves the stability.[125] They can entrap the light-weighed molecules, hemoglobin and genes during their bilayer membrane formation.[126]Additionally, some charged molecules or ionic amphiphiles such as dicetyl phosphate (DCP), phosphatidic , stearylamine (SA) and cetylpyridinium chloride are used as additives.[127]For the treatment of Glaucoma, various types of the niosomes have been papered and evaluated.[128]Although there are only few reports about use of nano-niosomes for antiviral activity, but it has potential to control drug delivery and release kinetics.[129]Vesicular nanocarrier based treatment of skin fungal infections have been studied by several research groups. These prepared formulations show several benefits like bioavailability enhancement of bioactivity, high skin permeation power, no side effects at application site, reduction in dosing frequency, and sustained drug release profile.[108] For the delivery of the natural drugs, niosomes are the one of the best alternatives, when compared to the traditional methods, as many problems associated with the natural drug delivery were overcome by using the niosomes type of the lipid based composites.[122]

Emulsomes: Loading lipophilic drugs in high concentrations is little challenging since the area occupied by hydrophilic internal core, in liposomes, is also large. If the internal core of liposome also contains some lipid loving material, then that may reinforce the loading capacity of lipophilic drug. Liposomal emulsion, concomitantly known as emulsome, can be the solution for this problem since its internal core is formulated with fats and triglycerides which stabilize the lipophilic drug in the form of emulsion while enhancing the bioavailability and minimizing toxic events. This solvent and surfactant free emulsomes comprise semi-solidified internal oil core which exhibits high efficiency in encapsulating water insoluble anti-tumor and antifungal drugs.[130]

Aquasomes: One of the recently developed system for the delivery of bioactive molecules like peptide, enzymes, protein, antigens and genes to specific sites are the three layered self-assembled nanocomposite platforms known as aquasomes. The spherical shaped (60-300 nm particle size) aquasomes are composed of a nanocrystalline solid ceramic core and a glassy polyhydroxyloligomeric surface coating upon which biochemically active molecules are adsorbed, with or without modification, via non-covalent interactions. The nanocrystalline solid ceramic core provides high degree of order and structural regularity to the aquasomes. Polyhydroxy oligomer coating offers water like background and safeguards fragile biochemically active molecules from dehydration, degradation and decomposition under physiological conditions. Broad range of payloads viz., poorly water soluble drugs, insulin, hemoglobin, oxygen, serratipeptidase have been effectively delivered via aquasomes.[131] Literature bears evidence that the bound biomolecules and drug payloads exhibit prolonged biological activity when delivered via aquasomes, while bound antigens have shown to evoke robust immune response. The drive behind the progress of aquasomes based bio-delivery vehicles is two-fold: Firstly, the carbohydrate layer of aquasomes acts as a protective coating which precludes any destructive denaturing collaboration between the therapeutic agent and the solid carriers. Secondly, aquasomes aids in preserving structural integrity and the ensuing pharmacological activity of the therapeutic payload, thereby increasing the probabilities of efficient unloading at the target site and promoting a better therapeutic effect.[132–134]

Conclusions

Major breakthroughs in nanomedicine has driven the investigations towards an enlarged use of liposome based composites for enhanced drug bioavailability, controlled release, aqueous solubility, external or internal stimuli-responsive drug unloading, site-specific targeting ability, and permeability in a step towards mitigating the adverse side effects of traditionally used lipid drug cargoes. The above attributes have been accomplished by the various lipid based excipients discussed throughout, thereby providing benefits to the temporal and/or spatial control of drug release. The sequential advancements in the LBDDS research initiated with overcoming the drawbacks associated with conventional liposomes such as easy degradation, inability to cross immunity system, low absorption or permeation by the development in stealth liposomes. To make the DDS more target specific and control the rate of drug release, advancements and developments in the design of stimuli-triggered LBDDS were accomplished. Further, the lipid conjugated drug composites opened up new avenues for larger molecules which were unable to cross the BBB. Incorporating nanoparticles in lipid based delivery formulations further expands the applications of the lipid composites in efficient drug delivery vehicles. Thus, the future of the LBDDS seems promising due to the profound interest of the formulation scientists. However, there are some barriers that need to be broken for bench-to-bedside translation of LBDDS. Firstly, following their uptake, the safety and toxicological profile of the various lipid-composite vehicles demands further investigation. Secondly, to preclude the premature release of the therapeutic drug, the structural and chemical stability during storage and post-administration needs to optimized. Finally, as far as industrialization is concerned, simplification of drug delivery design (biocompatibility, low cytotoxicity, lipid drug delivery formulation should provide full protection of the active agent), and synthesis (simpler methods, by adopting greener and bio-inspired processes) is the need of the hour for affordable LBDDS production at a large-scale.

References

[1]　K.K. Jain, Current Status and Future Prospects of Drug Delivery Systems, in: Drug Deliv. Syst., Academic, Humana Press, New York, NY, 2014: pp. 1–56. https://doi.org/10.1007/978-1-4939-0363-4_1

[2]　P. Pàmies, A. Stoddart, Materials for drug delivery, Nat. Mater. 12 (2013) 957–957. https://doi.org/10.1038/nmat3798

[3]　B.J. Boyd, T.H. Nguyen, A. Müllertz, Lipids in Oral Controlled Release Drug Delivery, in: Control. Release Oral Drug Deliv., Springer US, Boston, MA, 2011: pp. 299–327. https://doi.org/10.1007/978-1-4614-1004-1_15

[4]　A.T. Florence, A Short History of Controlled Drug Release and an Introduction, in: Control. Release Oral Drug Deliv., Springer US, Boston, MA, 2011: pp. 1–26. https://doi.org/10.1007/978-1-4614-1004-1_1

[5]　J. Reinholz, K. Landfester, V. Mailänder, The challenges of oral drug delivery via nanocarriers, Drug Deliv. 25 (2018) 1694. https://doi.org/10.1080/10717544.2018.1501119

[6]　H.D. Williams, N.L. Trevaskis, S.A. Charman, R.M. Shanker, W.N. Charman, C.W. Pouton, C.J.H. Porter, Strategies to address low drug solubility in discovery and development, Pharmacol. Rev. 65 (2013) 315–499. https://doi.org/10.1124/PR.112.005660

[7]　R.L. Carrier, L.A. Miller, I. Ahmed, The utility of cyclodextrins for enhancing oral bioavailability, J. Control. Release. 123 (2007) 78–99. https://doi.org/10.1016/J.JCONREL.2007.07.018

[8]　S. Kalepu, V. Nekkanti, Insoluble drug delivery strategies: Review of recent advances and business prospects, Acta Pharm. Sin. B. 5 (2015) 442–453. https://doi.org/10.1016/J.APSB.2015.07.003

[9]　S. Banerjee, A. Kundu, Lipid-drug conjugates: a potential nanocarrier system for oral drug delivery applications, DARU J. Pharm. Sci. 26 (2018) 65–75. https://doi.org/10.1007/s40199-018-0209-1

[10]　D. Irby, C. Du, F. Li, Lipid–drug conjugate for enhancing drug delivery, Mol. Pharm. 14 (2017) 1325–1338. https://doi.org/10.1021/acs.molpharmaceut.6b01027

[11] R.H. Müller, M. Radtke, S.A. Wissing, Solid lipid nanoparticles (SLN) and nanostructured lipid carriers (NLC) in cosmetic and dermatological preparations, Adv. Drug Deliv. Rev. 54 (2002) S131–S155. https://doi.org/10.1016/S0169-409X(02)00118-7

[12] M. Jadhao, P. Ahirkar, H. Kumar, R. Joshi, O.R. Meitei, S.K. Ghosh, Surfactant induced aggregation-disaggregation of photodynamic active chlorin e6 and its relevant interaction with DNA alkylating quinone in a biomimic micellar microenvironment, RSC Adv. 5 (2015). https://doi.org/10.1039/c5ra16181a

[13] M. Jadhao, R. Joshi, K. Ganorkar, S.K. Ghosh, Biomimetic systems trigger a benzothiazole based molecular switch to 'turn on' fluorescence, Spectrochim. Acta Part A Mol. Biomol. Spectrosc. 217 (2019) 197–205. https://doi.org/10.1016/J.SAA.2019.03.089

[14] S. Cikrikci, B. Mert, M.H. Oztop, Development of ph sensitive alginate/gum tragacanth based hydrogels for oral insulin delivery, J. Agric. Food Chem. 66 (2018) 11784–11796. https://doi.org/10.1021/acs.jafc.8b02525

[15] D.G. Wallace, J. Rosenblatt, Collagen gel systems for sustained delivery and tissue engineering, Adv. Drug Deliv. Rev. 55 (2003) 1631–1649. https://doi.org/10.1016/J.ADDR.2003.08.004

[16] B. Chatin, M. Mével, J. Devallière, L. Dallet, T. Haudebourg, P. Peuziat, T. Colombani, M. Berchel, O. Lambert, A. Edelman, B. Pitard, Liposome-based Formulation for intracellular delivery of functional proteins, Mol. Ther. Nucleic Acids. 4 (2015) 244. https://doi.org/10.1038/MTNA.2015.17

[17] J.C. Byeon, S.E. Lee, T.H. Kim, J. Bin Ahn, D.H. Kim, J.S. Choi, J.S. Park, Design of novel proliposome formulation for antioxidant peptide, glutathione with enhanced oral bioavailability and stability, Drug Deliv. 26 (2019) 216–225. https://doi.org/10.1080/10717544.2018.1551441

[18] B. Garnier, S. Tan, C. Gounou, A.R. Brisson, J. Laroche-Traineau, M.J. Jacobin-Valat, G. Clofent-Sanchez, Development of a Platform of antibody-presenting liposomes, Biointerphases. 7 (2012) 11. https://doi.org/10.1007/s13758-011-0011-9

[19] Wikipedia, Lipid polymorphism, (2018). https://en.wikipedia.org/wiki/Lipid_polymorphism

[20] H. Shrestha, R. Bala, S. Arora, Lipid-Based Drug Delivery Systems, J. Pharm. 2014 (2014) 10

[21]   S. Ahmed, A. Gull, M. Aqil, M. Danish Ansari, Y. Sultana, Poloxamer-407 thickened lipid colloidal system of agomelatine for brain targeting: Characterization, brain pharmacokinetic study and behavioral study on Wistar rats, Colloids Surfaces B Biointerfaces. 181(2019) 426-436. https://doi.org/10.1016/j.colsurfb.2019.05.016

[22]   T.N.Q. Phan, I. Shahzadi, A. Bernkop-Schnürch, Hydrophobic ion-pairs and lipid-based nanocarrier systems: The perfect match for delivery of BCS class 3 drugs, J. Control. Release. 304 (2019) 146–155. https://doi.org/10.1016/j.jconrel.2019.05.011

[23]   N.D.T. Le, P.H.L. Tran, B.J. Lee, T.T.D. Tran, Solid lipid particle-based tablets for buccal delivery: The role of solid lipid particles in drug release, J. Drug Deliv. Sci. Technol. 52 (2019) 96–102. https://doi.org/10.1016/j.jddst.2019.04.037

[24]   C.J.H. Porter, C.W. Pouton, J.F. Cuine, W.N. Charman, Enhancing intestinal drug solubilisation using lipid-based delivery systems, Adv. Drug Deliv. Rev. 60 (2008) 673–691. https://doi.org/10.1016/j.addr.2007.10.014

[25]   M.R.I. Shishir, L. Xie, C. Sun, X. Zheng, W. Chen, Advances in micro and nano-encapsulation of bioactive compounds using biopolymer and lipid-based transporters, Trends Food Sci. Technol. 78 (2018) 34–60. https://doi.org/10.1016/j.tifs.2018.05.018

[26]   M.S. Mufamadi, V. Pillay, Y.E. Choonara, L.C. Du Toit, G. Modi, D. Naidoo, V.M.K. Ndesendo, A Review on Composite Liposomal Technologies for Specialized Drug Delivery, J. Drug Deliv. 2011 (2011) 1–19. https://doi.org/10.1155/2011/939851

[27]   A. Jain, D. Thakur, G. Ghoshal, O.P. Katare, U.S. Shivhare, Characterization of microcapsulated beta-carotene formed by complex coacervation using casein and gum tragacanth, Int. J. Biol. Macromol. 87 (2016) 101–113. https://doi.org/10.1016/j.ijbiomac.2016.01.117

[28]   L. De Souza, D.A. Madalena, A.C. Pinheiro, J.A. Teixeira, A.A. Vicente, Ó.L. Ramos, Micro- and nano bio-based delivery systems for food applications : In vitro behavior, 243 (2017) 23–45. https://doi.org/10.1016/j.cis.2017.02.010

[29]   X. Xu, M.A. Khan, D.J. Burgess, Predicting hydrophilic drug encapsulation inside unilamellar liposomes, Int. J. Pharm. 423 (2012) 410–418. https://doi.org/10.1016/j.ijpharm.2011.12.019

[30]   S.M. Jafari, D.J. McClements, Nanotechnology Approaches for Increasing Nutrient Bioavailability, 1st ed., Elsevier Inc., 2017. https://doi.org/10.1016/bs.afnr.2016.12.008

[31]   G. Liu, W. Huang, O. Babii, X. Gong, Z. Tian, J. Yang, Y. Wang, R.L. Jacobs, V. Donna, A. Lavasanifar, L. Chen, Novel protein–lipid composite nanoparticles with an

inner aqueous compartment as delivery systems of hydrophilic nutraceutical compounds, Nanoscale. 10 (2018) 10629–10640. https://doi.org/10.1039/C8NR01009A

[32]   R.K. Harwansh, R. Deshmukh, M.A. Rahman, Nanoemulsion: Promising nanocarrier system for delivery of herbal bioactives, J. Drug Deliv. Sci. Technol. 51 (2019) 224–233. https://doi.org/10.1016/j.jddst.2019.03.006

[33]   Y.S.R. Elnaggar, S. Omran, H.A. Hazzah, O.Y. Abdallah, Anionic versus cationic bilosomes as oral nanocarriers for enhanced delivery of the hydrophilic drug risedronate, Int. J. Pharm. 564 (2019) 410–425. https://doi.org/10.1016/j.ijpharm.2019.04.069

[34]   B.D. Buddy, D. Ratner, Biomaterials science : An introduction to materials in medicine, Academic Press, 2013

[35]   A. George, P.A. Shah, P.S. Shrivastav, Natural biodegradable polymers based nano-formulations for drug delivery: A review, Int. J. Pharm. 561 (2019) 244–264. https://doi.org/10.1016/j.ijpharm.2019.03.011

[36]   L.N.M. Ribeiro, A.C.S. Alcântara, G.H. Rodrigues da Silva, M. Franz-Montan, S.V.G. Nista, S.R. Castro, V.M. Couto, V.A. Guilherme, E. de Paula, Advances in hybrid polymer-based materials for sustained drug release, Int. J. Polym. Sci. 2017 (2017) 1–16. https://doi.org/10.1155/2017/1231464

[37]   Y. Cheng, T. Zou, M. Dai, X.Y. He, N. Peng, K. Wu, X.Q. Wang, C.Y. Liao, Y. Liu, Doxorubicin loaded tumor-triggered targeting ammonium bicarbonate liposomes for tumor-specific drug delivery, Colloids Surfaces B Biointerfaces. 178 (2019) 263–268. https://doi.org/10.1016/j.colsurfb.2019.03.002

[38]   Y. Fang, M. Vadlamudi, Y. Huang, X. Guo, Lipid-Coated, pH-Sensitive Magnesium Phosphate Particles for Intracellular Protein Delivery, Pharm. Res. 36 (2019) 81. https://doi.org/10.1007/s11095-019-2607-6

[39]   Z. Zhao, W. Yao, N. Wang, C. Liu, H. Zhou, H. Chen, W. Qiao, Synthesis and evaluation of mono- and multi-hydroxyl low toxicity pH-sensitive cationic lipids for drug delivery, Eur. J. Pharm. Sci. 133 (2019) 69–78. https://doi.org/10.1016/j.ejps.2019.03.018

[40]   M. Tang, D. Svirskis, E. Leung, M. Kanamala, H. Wang, Z. Wu, Can intracellular drug delivery using hyaluronic acid functionalised pH-sensitive liposomes overcome gemcitabine resistance in pancreatic cancer, J. Control. Release. 305 (2019) 89–100. https://doi.org/10.1016/j.jconrel.2019.05.018

[41] Y. Yao, T. Wang, Y. Liu, N. Zhang, Co-delivery of sorafenib and VEGF-siRNA via pH-sensitive liposomes for the synergistic treatment of hepatocellular carcinoma, Artif. Cells Nanomedicine Biotechnol. 47 (2019) 1374–1383. https://doi.org/10.1080/21691401.2019.1596943

[42] M. Chen, F. Song, Y. Liu, J. Tian, C. Liu, R. Li, Q. Zhang, A dual pH-sensitive liposomal system with charge-reversal and NO generation for overcoming multidrug resistance in cancer, Nanoscale. 11 (2019) 3814–3826. https://doi.org/10.1039/c8nr06218h

[43] G.N. Pawar, N.N. Parayath, A.L. Nocera, B.S. Bleier, M.A. Id, Direct CNS delivery of proteins using thermosensitive liposome-in-gel carrier by heterotopic mucosal engrafting, PLOS ONE. (2018) 1–15

[44] S. Akthar, R.J. Kok, T. Lammers, G. Storm, Influence of cholesterol inclusion on the doxorubicin release characteristics of lysolipid-based thermosensitive liposomes, Int. J. Pharm. (2017). https://doi.org/10.1016/j.ijpharm.2017.11.002

[45] D. Cao, X. Zhang, Y. Luo, H. Wu, X. Ke, F. Group, Liposomal doxorubicin loaded PLGA-PEG-PLGA based thermogel for sustained local drug delivery for the treatment of breast cancer local drug delivery for the treatment of breast cancer, Artif. Cells, Nanomedicine, Biotechnol. 47 (2019) 181–191. https://doi.org/10.1080/21691401.2018.1548470

[46] Y. Mo, H. Du, B. Chen, D. Liu, Q. Yin, Y. Yan, Z. Wang, F. Wan, T. Qi, Y. Wang, Q. Zhang, Y. Wang, Quick-responsive polymer-based thermosensitive liposomes for controlled doxorubicin release and chemotherapy, ACS Biomater. Sci. Eng. 5 (2019) 2316–2329. https://doi.org/10.1021/acsbiomaterials.9b00343

[47] Y.J. Lu, E.Y. Chuang, Y.H. Cheng, T.S. Anilkumar, H.A. Chen, J.P. Chen, Thermosensitive magnetic liposomes for alternating magnetic field-inducible drug delivery in dual targeted brain tumor chemotherapy, Chem. Eng. J. 373 (2019) 720–733. https://doi.org/10.1016/j.cej.2019.05.055

[48] X. Wang, R. Yang, C. Yuan, Y. An, Q. Tang, D. Chen, Preparation of Folic Acid-Targeted Temperature-Sensitive Magnetoliposomes and their Antitumor Effects In Vitro and In Vivo, Targ Oncol. 13 (2018) 481

[49] M. Ayubi, M. Karimi, S. Abdpour, K. Rostamizadeh, M. Parsa, M. Zamani, A. Saedi, Magnetic nanoparticles decorated with PEGylated curcumin as dual targeted drug delivery: Synthesis, toxicity and biocompatibility study, Mater. Sci. Eng. C. 104 (2019) 109810. https://doi.org/10.1016/j.msec.2019.109810

[50]   M. Suñé-Pou, M.J. Limeres, I. Nofrerias, A. Nardi-Ricart, S. Prieto-Sánchez, Y. El-Yousfi, P. Pérez-Lozano, E. García-Montoya, M. Miñarro-Carmona, J.R. Ticó, C. Hernández-Munain, C. Suñé, J.M. Suñé-Negre, Improved synthesis and characterization of cholesteryl oleate-loaded cationic solid lipid nanoparticles with high transfection efficiency for gene therapy applications, Colloids Surfaces B Biointerfaces. 180 (2019) 159–167. https://doi.org/10.1016/j.colsurfb.2019.04.037

[51]   H.M. Eid, M.H. Elkomy, S.F. El Menshawe, H.F. Salem, Development, optimization, and in vitro/in vivo characterization of enhanced lipid nanoparticles for ocular delivery of ofloxacin: the influence of pegylation and chitosan coating, AAPS PharmSciTech. 20 (2019) 1–14. https://doi.org/10.1208/s12249-019-1371-6

[52]   I. Ahmad, J. Pandit, Y. Sultana, A.K. Mishra, P.P. Hazari, M. Aqil, Optimization by design of etoposide loaded solid lipid nanoparticles for ocular delivery: Characterization, pharmacokinetic and deposition study, Mater. Sci. Eng. C. 100 (2019) 959–970. https://doi.org/10.1016/j.msec.2019.03.060

[53]   K.Y. Janga, A. Tatke, N. Dudhipala, S.P. Balguri, M.M. Ibrahim, D.N. Maria, M.M. Jablonski, S. Majumdar, Gellan gum based sol-to-gel transforming system of natamycin transfersomes improves topical ocular delivery, J. Pharmacol. Exp. Ther. (2019) jpet.119.256446. https://doi.org/10.1124/jpet.119.256446

[54]   G.S. El-feky, M.M. El-naa, A.A. Mahmoud, Journal of Drug Delivery Science and Technology Flexible nano-sized lipid vesicles for the transdermal delivery of colchicine ; in vitro / in vivo investigation, J. Drug Deliv. Sci. Technol. 49 (2019) 24–34. https://doi.org/10.1016/j.jddst.2018.10.036

[55]   M.K. Shamshiri, M.K. Shahraky, F. Rahimi, Lecithin soybean phospholipid nano transfersomes as potential carriers for transdermal delivery of the human growth hormone, Journal of Cellular Biochemistry. 120 (2019) 9023-9033. https://doi.org/10.1002/jcb.28176

[56]   S. Rajalakshmi, N. Vyawahare, A. Pawar, P. Mahaparale, B. Chellampillai, Current development in novel drug delivery systems of bioactive molecule plumbagin, Artif. Cells, Nanomedicine Biotechnol. 46 (2018) 209–218. https://doi.org/10.1080/21691401.2017.1417865

[57]   G.M. El-Zaafarany, M.E. Soliman, S. Mansour, M. Cespi, G.F. Palmieri, L. Illum, L. Casettari, G.A.S. Awad, A tailored thermosensitive PLGA-PEG-PLGA/emulsomes composite for enhanced oxcarbazepine brain delivery via the nasal route, Pharmaceutics. 10 (2018). https://doi.org/10.3390/pharmaceutics10040217

[58]   G.M. El-Zaafarany, M.E. Soliman, S. Mansour, G.A.S. Awad, Identifying lipidic emulsomes for improved oxcarbazepine brain targeting: In vitro and rat in vivo studies, Int. J. Pharm. 503 (2016) 127–140. https://doi.org/10.1016/j.ijpharm.2016.02.038

[59]   D. Sawant, P.M. Dandagi, A.P. Gadad, Formulation and evaluation of sparfloxacin emulsomes-loaded thermosensitive in situ gel for ophthalmic delivery, J. Sol-Gel Sci. Technol. 77 (2016) 654–665. https://doi.org/10.1007/s10971-015-3897-8

[60]   L. Inchaurraga, A.L. Martínez-López, M. Abdulkarim, M. Gumbleton, G. Quincoces, I. Peñuelas, N. Martin-Arbella, J.M. Irache, Modulation of the fate of zein nanoparticles by their coating with a Gantrez® AN-thiamine polymer conjugate, Int. J. Pharm. X. 1 (2019) 100006. https://doi.org/10.1016/j.ijpx.2019.100006

[61]   M.J. Costa, J. Kudaravalli, J.T. Ma, W.H. Ho, K. Delaria, C. Holz, A. Stauffer, A.G. Chunyk, Q. Zong, E. Blasi, B. Buetow, T.T. Tran, K. Lindquist, M. Dorywalska, A. Rajpal, D.L. Shelton, P. Strop, S.H. Liu, Optimal design, anti-tumour efficacy and tolerability of anti-CXCR4 antibody drug conjugates, Sci. Rep. 9 (2019) 1–19. https://doi.org/10.1038/s41598-019-38745-x

[62]   N. Plenagl, L. Duse, B.S. Seitz, N. Goergen, S.R. Pinnapireddy, J. Jedelska, J. Brüßler, U. Bakowsky, Photodynamic therapy - hypericin tetraether liposome conjugates and their antitumor and antiangiogenic activity, Drug Deliv. 26 (2019) 23–33. https://doi.org/10.1080/10717544.2018.1531954

[63]   L.P. Mendes, C. Sarisozen, E. Luther, J. Pan, V.P. Torchilin, Surface-engineered polyethyleneimine-modified liposomes as novel carrier of siRNA and chemotherapeutics for combination treatment of drug-resistant cancers, Drug Deliv. 26 (2019) 443–458. https://doi.org/10.1080/10717544.2019.1574935

[64]   K. Kristensen, T.B. Engel, A. Stensballe, J.B. Simonsen, T.L. Andresen, The hard protein corona of stealth liposomes is sparse, J. Control. Release. 307 (2019) 1–15. https://doi.org/10.1016/j.jconrel.2019.05.042

[65]   N.S. Awad, V. Paul, M.H. Al-Sayah, G.A. Husseini, Ultrasonically controlled albumin-conjugated liposomes for breast cancer therapy, Artif. Cells, Nanomedicine Biotechnol. 47 (2019) 705–714. https://doi.org/10.1080/21691401.2019.1573175

[66]   M.H. Jang, C.H. Kim, H.Y. Yoon, S.W. Sung, M.S. Goh, E.S. Lee, D.J. Shin, Y.W. Choi, Steric stabilization of RIPL peptide-conjugated liposomes and in vitro assessment, J. Pharm. Investig. 49 (2019) 115–125. https://doi.org/10.1007/s40005-018-0392-6

[67]   D.D. Lasic, D. Needham, The "Stealth" Liposome: A Prototypical Biomaterial, Chem. Rev. 95 (1995) 2601–2628. https://doi.org/10.1021/cr00040a001

[68]   B. C̆eh, M. Winterhalter, P.M. Frederik, J.J. Vallner, D.D. Lasic, Stealth® liposomes: from theory to product, Adv. Drug Deliv. Rev. 24 (1997) 165–177. https://doi.org/10.1016/S0169-409X(96)00456-5

[69]   J. Liu, C. Detrembleur, S. Mornet, C. Jérôme, E. Duguet, Design of hybrid nanovehicles for remotely triggered drug release: An overview, J. Mater. Chem. B. 3 (2015) 6117–6147. https://doi.org/10.1039/C5TB00664C

[70]   M. Karimi, A. Ghasemi, P. Sahandi Zangabad, R. Rahighi, S.M. Moosavi Basri, H. Mirshekari, M. Amiri, Z. Shafaei Pishabad, A. Aslani, M. Bozorgomid, D. Ghosh, A. Beyzavi, A. Vaseghi, A.R. Aref, L. Haghani, S. Bahrami, M.R. Hamblin, Smart micro/nanoparticles in stimulus-responsive drug/gene delivery systems, Chem. Soc. Rev. 45 (2016) 1457–1501. https://doi.org/10.1039/C5CS00798D

[71]   Y. Matsuzaki, Y. Hamasaki, S.I. Said, pH-Sensitive Liposomes : Possible Clinical Implications, Science. 210 (1980) 1253–1255

[72]   S.R. Paliwal, R. Paliwal, S.P. Vyas, A review of mechanistic insight and application of pH-sensitive liposomes in drug delivery, 7544 (2014) 1–12. https://doi.org/10.3109/10717544.2014.882469

[73]   S. Mallick, J.S. Choi, Liposomes: Versatile and Biocompatible Nanovesicles for Efficient Biomolecules Delivery, J. Nanosci. Nanotechnol. 14 (2014) 755–765. https://doi.org/10.1166/jnn.2014.9080

[74]   S. Mura, J. Nicolas, P. Couvreur, Stimuli-responsive nanocarriers for drug delivery, Nat. Publ. Gr. 12 (2013) 991–1003. https://doi.org/10.1038/nmat3776

[75]   S. Ganta, H. Devalapally, A. Shahiwala, M. Amiji, A review of stimuli-responsive nanocarriers for drug and gene delivery, J. Control. Release. 126 (2008) 187–204. https://doi.org/10.1016/j.jconrel.2007.12.017

[76]   H. Bi, J. Xue, H. Jiang, S. Gao, D. Yang, Y. Fang, K. Shi, Current developments in drug delivery with thermosensitive liposomes, Asian J. Pharm. Sci. (2018). https://doi.org/10.1016/J.AJPS.2018.07.006

[77]   G. Bozzuto, A. Molinari, Liposomes as nanomedical devices, Int. J. Nanomedicine. 10 (2015) 975–999. https://doi.org/10.2147/IJN.S68861

[78]   U. States, Thermosensitive Liposomes for Image-Guided Drug Delivery, 1st ed., Elsevier Inc., 2018. https://doi.org/10.1016/bs.acr.2018.04.004

[79]   B. Kneidl, M. Peller, G. Winter, L.H. Lindner, M. Hossann, Thermosensitive liposomal drug delivery systems : State of the art review, Int J Nanomedicine. (2014) 4387–4398

[80]   W. Rawicz, K.C. Olbrich, T. McIntosh, D. Needham, E. Evans, Effect of Chain Length and Unsaturation on Elasticity of Lipid Bilayers, Biophys. J. 79 (2000) 328–339. https://doi.org/10.1016/S0006-3495(00)76295-3

[81]   J. Massiot, V. Rosilio, A. Makky, Photo-triggerable liposomal drug delivery systems: from simple porphyrin insertion in the lipid bilayer towards supramolecular assemblies of lipid–porphyrin conjugates, J. Mater. Chem. B. 7 (2019) 1805–1823. https://doi.org/10.1039/C9TB00015A

[82]   S.J. Leung, M. Romanowski, Light-Activated Content Release from Liposomes, Theranostics. 2 (2012) 1020–1036. https://doi.org/10.7150/thno.4847

[83]   B.S. Pattni, V. V. Chupin, V.P. Torchilin, New Developments in Liposomal Drug Delivery, Chem. Rev. 115 (2015) 10938–10966. https://doi.org/10.1021/acs.chemrev.5b00046

[84]   A. Yavlovich, B. Smith, K. Gupta, R. Blumenthal, A. Puri, Light-sensitive lipid-based nanoparticles for drug delivery: design principles and future considerations for biological applications., Mol. Membr. Biol. 27 (2010) 364–81. https://doi.org/10.3109/09687688.2010.507788

[85]   D. Conceição, D. Ferreira, L. Ferreira, D.S. Conceição, D.P. Ferreira, L.F.V. Ferreira, Photochemistry and Cytotoxicity Evaluation of Heptamethinecyanine Near Infrared (NIR) Dyes, Int. J. Mol. Sci. 14 (2013) 18557–18571. https://doi.org/10.3390/ijms140918557

[86]   B. Chandra, S. Mallik, D.K. Srivastava, Design of photocleavable lipids and their application in liposomal "uncorking," Chem. Commun. (2005) 3021. https://doi.org/10.1039/b503423j

[87]   B. Chandra, R. Subramaniam, S. Mallik, D.K. Srivastava, Formulation of photocleavable liposomes and the mechanism of their content release, Org. Biomol. Chem. 4 (2006) 1730. https://doi.org/10.1039/b518359f

[88]   S.L. Regen, A. Singh, G. Oehme, M. Singh, Polymerized phosphatidyl choline vesicles. Stabilized and controllable time-release carriers., Biochem. Biophys. Res. Commun. 101 (1981) 131–6. https://doi.org/10.1016/s0006-291x(81)80020-4

[89]   A. Yavlovich, A. Singh, S. Tarasov, J. Capala, R. Blumenthal, A. Puri, Design of liposomes containing photopolymerizable phospholipids for triggered release of

contents, J. Therm. Anal. Calorim. 98 (2009) 97–104. https://doi.org/10.1007/s10973-009-0228-8

[90] P. Shum, J.M. Kim, D.H. Thompson, Phototriggering of liposomal drug delivery systems., Adv. Drug Deliv. Rev. 53 (2001) 273–84

[91] Z. Li, Y. Wan, A.G. Kutateladze, Dithiane-based photolabile amphiphiles: toward photolabile liposomes1,2, Langmuir. 19 (2003) 6381–6391. https://doi.org/10.1021/LA034188M

[92] M. Mathiyazhakan, C. Wiraja, C. Xu, A Concise review of gold nanoparticles-based photo-responsive liposomes for controlled drug delivery, Nano-Micro Lett. 10 (2018) 10. https://doi.org/10.1007/s40820-017-0166-0

[93] H. Basoglu, M.D. Bilgin, M.M. Demir, Protoporphyrin IX-loaded magnetoliposomes as a potential drug delivery system for photodynamic therapy: Fabrication, characterization and in vitro study, Photodiagnosis Photodyn. Ther. 13 (2016) 81–90. https://doi.org/10.1016/j.pdpdt.2015.12.010

[94] A. Deshpande, M. Mohamed, S.B. Daftardar, M. Patel, S.H.S. Boddu, J. Nesamony, Solid Lipid nanoparticles in drug delivery: opportunities and challenges, emerg. nanotechnologies diagnostics, Drug Deliv. Med. Devices. (2017) 291–330. https://doi.org/10.1016/B978-0-323-42978-8.00012-7

[95] R.H. Müller, K. Mäder, S. Gohla, Solid lipid nanoparticles (SLN) for controlled drug delivery – A review of the state of the art, Eur. J. Pharm. Biopharm. 50 (2000) 161–177. https://doi.org/10.1016/S0939-6411(00)00087-4

[96] S. Mukherjee, S. Ray, R.S. Thakur, Solid lipid nanoparticles: a modern formulation approach in drug delivery system, Indian J. Pharm. Sci. 71 (2009) 349. https://doi.org/10.4103/0250-474X.57282

[97] N. Dudhipala, K. Veerabrahma, Pharmacokinetic and pharmacodynamic studies of nisoldipine-loaded solid lipid nanoparticles developed by central composite design, Drug Dev. Ind. Pharm. 41 (2015) 1968–1977. https://doi.org/10.3109/03639045.2015.1024685

[98] R.H. Müller, U. Alexiev, P. Sinambela, C.M. Keck, Nanostructured Lipid Carriers (NLC): The Second Generation of Solid Lipid Nanoparticles, in: Percutaneous Penetration Enhanc. Chem. Methods Penetration Enhanc., Springer Berlin Heidelberg, Berlin, Heidelberg, 2016: pp. 161–185. https://doi.org/10.1007/978-3-662-47862-2_11

[99] J.Y. Fang, C.L. Fang, C.H. Liu, Y.H. Su, Lipid nanoparticles as vehicles for topical psoralen delivery: Solid lipid nanoparticles (SLN) versus nanostructured lipid

carriers (NLC), Eur. J. Pharm. Biopharm. 70 (2008) 633–640.
https://doi.org/10.1016/J.EJPB.2008.05.008

[100] F. Tamjidi, M. Shahedi, J. Varshosaz, A. Nasirpour, Nanostructured lipid carriers
(NLC): A potential delivery system for bioactive food molecules, Innov. Food Sci.
Emerg. Technol. 19 (2013) 29–43. https://doi.org/10.1016/J.IFSET.2013.03.002

[101] B. Debelec-Butuner, M. Kotmakci, E. Oner, G. Ozduman, A.G. Kantarci,
Nutlin3a-loaded nanoparticles show enhanced apoptotic activity on prostate cancer
cells, Mol. Biotechnol. (2019). https://doi.org/10.1007/s12033-019-00178-2

[102] X. Liu, Q. Zhao, Long-term anesthetic analgesic effects: Comparison of tetracaine
loaded polymeric nanoparticles, solid lipid nanoparticles, and nanostructured lipid
carriers in vitro and in vivo, Biomed. Pharmacother. 117 (2019) 109057.
https://doi.org/10.1016/j.biopha.2019.109057

[103] M. Üner, G. Yener, M. Ergüven, Design of colloidal drug carriers of celecoxib for
use in treatment of breast cancer and leukemia, Mater. Sci. Eng. C. 103 (2019)
109874. https://doi.org/10.1016/j.msec.2019.109874

[104] M. Patel, V. Mundada, K. Sawant, Enhanced intestinal absorption of asenapine
maleate by fabricating solid lipid nanoparticles using TPGS: Elucidation of transport
mechanism, permeability across Caco-2 cell line and in vivo pharmacokinetic studies,
Artif. Cells, Nanomedicine, Biotechnol. 47 (2019) 144–153.
https://doi.org/10.1080/21691401.2018.1546186

[105] R. Bharadwaj, B.P. Sahu, J. Haloi, D. Laloo, P. Barooah, C. Keppen, M. Deka, S.
Medhi, Combinatorial therapeutic approach for treatment of oral squamous cell
carcinoma, Artif. Cells, Nanomedicine Biotechnol. 47 (2019) 572–585.
https://doi.org/10.1080/21691401.2019.1573176

[106] R.H. Muller, C.M. Keck, Challenges and solutions for the delivery of biotech
drugs – a review of drug nanocrystal technology and lipid nanoparticles, J. Biotechnol.
113 (2004) 151–170. https://doi.org/10.1016/J.JBIOTEC.2004.06.007

[107] P. Adhikari, P. Pal, A.K. Das, S. Ray, A. Bhattacharjee, B. Mazumder, Nano lipid-
drug conjugate: An integrated review, Int. J. Pharm. 529 (2017) 629–641.
https://doi.org/10.1016/J.IJPHARM.2017.07.039

[108] S. Verma, P. Utreja, Vesicular nanocarrier based treatment of skin fungal
infections: Potential and emerging trends in nanoscale pharmacotherapy, Asian J.
Pharm. Sci. 14 (2019) 117–129. https://doi.org/10.1016/j.ajps.2018.05.007

[109] H.A. Benson, Transfersomes for transdermal drug delivery, Expert Opin. Drug Deliv. 3 (2006) 727–737. https://doi.org/10.1517/17425247.3.6.727

[110] R. Rajan, S. Jose, V.P.B. Mukund, D.T. Vasudevan, Transferosomes - A vesicular transdermal delivery system for enhanced drug permeation, J. Adv. Pharm. Technol. Res. 2 (2011) 138–43. https://doi.org/10.4103/2231-4040.85524

[111] D.A.Y. Pawar, Transfersome: A novel technique which improves transdermal permeability, Asian J. Pharm. 10 (2016) 425–436. https://doi.org/10.22377/AJP.V10I04.875

[112] G. Cevc, G. Blume, Lipid vesicles penetrate into intact skin owing to the transdermal osmotic gradients and hydration force, Biochim. Biophys. Acta - Biomembr. 1104 (1992) 226–232. https://doi.org/10.1016/0005-2736(92)90154-E

[113] B. Bhasin, S.P. Patel, V.L.M. Road, An overview of transfersomal drug delivery bhavya bhasin and Vaishali Y. Londhe SVKM'S NMIMS , Shobhaben Pratapbhai Patel School of Pharmacy and Technology Management, Mumbai - 400056, Maharashtra, India., 9 (2018) 2175–2184. https://doi.org/10.13040/IJPSR.0975-8232.9(6).2175-84

[114] B. Bhasin, S.P. Patel, V.L.M. Road, An overview of transfersomal drug delivery, Int. J. Pharm. Sci. Res. 9 (2018) 2175–2184. https://doi.org/10.13040/IJPSR.0975-8232.9(6).2175-84

[115] A. Waheed, M. Aqil, A. Ahad, S.S. Imam, T. Moolakkadath, Z. Iqbal, A. Ali, AC SC, J. Drug Deliv. Sci. Technol. (2019). https://doi.org/10.1016/j.jddst.2019.05.019

[116] H. Song, H. Li, Y. Meng, Y. Zhang, N. Zhang, W. Zheng, Enhanced transdermal permeability and drug deposition of rheumatoid arthritis via sinomenine hydrochloride-loaded antioxidant surface transethosome, Int J Nanomedicine. 14 (2019) 3177–3188

[117] M.M. Omar, O.A. Hasan, A.M. El Sisi, Preparation and optimization of lidocaine transferosomal gel containing permeation enhancers : A promising approach for enhancement of skin permeation, Int J Nanomedicine. 14 (2019) 1551–1562

[118] P. Chaurasiya, E. Ganju, N. Upmanyu, S.K. Ray, P. Jain, Transfersomes: a novel technique for transdermal drug delivery, J. Drug Deliv. Ther. 9 (2019) 279–285. https://doi.org/10.22270/jddt.v9i1.2198

[119] A. Sankhyan, P. Pawar, Recent trends in niosome as vesicular drug delivery system, J. Appl. Pharm. Sci. 2 (2012) 20–32. https://doi.org/10.7324/JAPS.2012.2625

[120] I.F. Uchegbu, S.P. Vyas, Non-ionic surfactant based vesicles (niosomes) in drug delivery, Int. J. Pharm. 172 (1998) 33–70. https://doi.org/10.1016/S0378-5173(98)00169-0

[121] M.J. Choi, H.I. Maibach, Liposomes and niosomes as topical drug delivery systems, Skin Pharmacol. Physiol. 18 (2005) 209–219. https://doi.org/10.1159/000086666

[122] M. Gharbavi, J. Amani, H. Kheiri-Manjili, H. Danafar, A. Sharafi, Niosome: A promising nanocarrier for natural drug delivery through blood-brain barrier, Adv. Pharmacol. Sci. 2018 (2018). https://doi.org/10.1155/2018/6847971

[123] G.P. Kumar, P. Rajeshwarrao, Nonionic surfactant vesicular systems for effective drug delivery—an overview, Acta Pharm. Sin. B. 1 (2011) 208–219. https://doi.org/10.1016/J.APSB.2011.09.002

[124] T. Liu, R. Guo, Preparation of a highly stable niosome and its hydrotrope-solubilization action to drugs, Langmuir. 21-24 (2005) 11034-11039. https://doi.org/10.1021/LA051868B

[125] R. Rajera, K. Nagpal, S.K. Singh, D.N. Mishra, Niosomes: a controlled and novel drug delivery system., Biol. Pharm. Bull. 34 (2011) 945–53

[126] G. Amoabediny, F. Haghiralsadat, S. Naderinezhad, M.N. Helder, E. Akhoundi Kharanaghi, J. Mohammadnejad Arough, B. Zandieh-Doulabi, Overview of preparation methods of polymeric and lipid-based (niosome, solid lipid, liposome) nanoparticles: A comprehensive review, Int. J. Polym. Mater. Polym. Biomater. 67 (2018) 383–400. https://doi.org/10.1080/00914037.2017.1332623

[127] X. Ge, M. Wei, S. He, W.-E. Yuan, Advances of non-ionic surfactant vesicles (niosomes) and their application in drug delivery, Pharmaceutics. 11 (2019) 55. https://doi.org/10.3390/pharmaceutics11020055

[128] K.S. Yadav, R. Rajpurohit, S. Sharma, Glaucoma: Current treatment and impact of advanced drug delivery systems, Life Sci. 221 (2019) 362–376. https://doi.org/10.1016/j.lfs.2019.02.029

[129] R.G. Kerry, S. Malik, Y.T. Redda, S. Sahoo, J.K. Patra, S. Majhi, Nano-based approach to combat emerging viral (NIPAH virus) infection, Nanomedicine Nanotechnology Biol. Med. 18 (2019) 196–220. https://doi.org/10.1016/j.nano.2019.03.004

[130] A. Pal, S. Gupta, A. Jaiswal, A. Dube, S.P. Vyas, Development and evaluation of tripalmitin emulsomes for the treatment of experimental visceral leishmaniasis, J. Liposome Res. 22 (2012) 62–71. https://doi.org/10.3109/08982104.2011.592495

[131] N. Kossovsky, Biomolecular Delivery Using Coated Nanocrystalline Ceramics (Aquasomes), Nanotechnology. 626 (1996) 334–350. https://doi.org/10.1021/bk-1996-0622.ch023

[132] S. Banerjee, K.K. Sen, Aquasomes: A novel nanoparticulate drug carrier, J. Drug Deliv. Sci. Technol. 43 (2018) 446–452. https://doi.org/10.1016/J.JDDST.2017.11.011

[133] M.S. Umashankar, R.K. Sachdeva, M. Gulati, Aquasomes: A promising carrier for peptides and protein delivery, Nanomedicine Nanotechnology Biol. Med. 6 (2010) 419–426. https://doi.org/10.1016/J.NANO.2009.11.002

[134] S.S. Jain, P.S. Jagtap, N.M. Dand, K.R. Jadhav, V.J. Kadam, Aquasomes: A novel drug carrier, J. Appl. Pharm. Sci. 02 (2012) 184–192

Keyword Index

About the Editors

Dr. Amir Al-Ahmed is working as a Research Scientist-II (Associate Professor) in the Center of Research Excellence in Renewable Energy, at King Fahd University of Petroleum & Minerals (KFUPM), Saudi Arabia. He graduated in chemistry from Aligarh Muslim University (AMU), India. He obtained his M.Phil (2001) and Ph.D. (2003) degree in Applied Chemistry on conducting polymer based composites and its applications, from the Zakir Hussain College of Engineering and Technology, AMU, India. During his postdoctoral research activity, he worked on different multi-disciplinary project in South Africa and Saudi Arabia such as biological and chemical sensor, energy storage and conversion and also on CO2 reduction. Throughout his academic career, he has gained extensive experience in materials chemistry and electrocatalysis for applications in energy conversion, storage, sensors and membranes. Currently, he is involved in several multidisciplinary research projects, funded by Saudi national research programs (thin film solar cells) and Saudi Aramco. Dr. Amir has on his credit a good number of articles published in international scientific journals, conferences and as book chapters. He has edited six books (publisher: Trans Tech Publication, Switzerland) and he is in the process of editing and writing another two books with Springer and Elsevier. He is also the Editor-in-Chief of an international journal "Nano Hybrids and Composites" along with Professor Y.H. Kim.

Dr. Inamuddin is currently working as Assistant Professor in the Chemistry Department, Faculty of Science, King Abdulaziz University, Jeddah, Saudi Arabia. He is a permanent faculty member (Assistant Professor) at the Department of Applied Chemistry, Aligarh Muslim University, Aligarh, India. He obtained Master of Science degree in Organic Chemistry from Chaudhary Charan Singh (CCS) University, Meerut, India, in 2002. He received his Master of Philosophy and Doctor of Philosophy degrees in Applied Chemistry from Aligarh Muslim University (AMU), India, in 2004 and 2007, respectively. He has extensive research experience in multidisciplinary fields of Analytical Chemistry, Materials Chemistry, and Electrochemistry and, more specifically, Renewable Energy and Environment. He has worked on different research projects as project fellow and senior research fellow funded by University Grants Commission (UGC), Government of India, and Council of Scientific and Industrial Research (CSIR), Government of India. He has received Fast Track Young Scientist Award from the Department of Science and Technology, India, to work in the area of bending actuators and artificial muscles. He has completed four major research projects sanctioned by University Grant Commission, Department of Science and Technology, Council of

Scientific and Industrial Research, and Council of Science and Technology, India. He has published 147 research articles in international journals of repute and eighteen book chapters in knowledge-based book editions published by renowned international publishers. He has published 60 edited books with Springer (U.K.), Elsevier, Nova Science Publishers, Inc. (U.S.A.), CRC Press Taylor & Francis Asia Pacific, Trans Tech Publications Ltd. (Switzerland), IntechOpen Limited (U.K.), and Materials Research Forum LLC (U.S.A). He is a member of various journals' editorial boards. He is also serving as Associate Editor for journals (Environmental Chemistry Letter, Applied Water Science and Euro-Mediterranean Journal for Environmental Integration, Springer-Nature), Frontiers Section Editor (Current Analytical Chemistry, Bentham Science Publishers), Editorial Board Member (Scientific Reports-Nature), Editor (Eurasian Journal of Analytical Chemistry), and Review Editor (Frontiers in Chemistry, Frontiers, U.K.) He is also guest-editing various special thematic special issues to the journals of Elsevier, Bentham Science Publishers, and John Wiley & Sons, Inc. He has attended as well as chaired sessions in various international and national conferences. He has worked as a Postdoctoral Fellow, leading a research team at the Creative Research Initiative Center for Bio-Artificial Muscle, Hanyang University, South Korea, in the field of renewable energy, especially biofuel cells. He has also worked as a Postdoctoral Fellow at the Center of Research Excellence in Renewable Energy, King Fahd University of Petroleum and Minerals, Saudi Arabia, in the field of polymer electrolyte membrane fuel cells and computational fluid dynamics of polymer electrolyte membrane fuel cells. He is a life member of the Journal of the Indian Chemical Society. His research interest includes ion exchange materials, a sensor for heavy metal ions, biofuel cells, supercapacitors and bending actuators.

www.ingramcontent.com/pod-product-compliance
Lightning Source LLC
Chambersburg PA
CBHW071212210326
41597CB00016B/1779